MOSFETs

By

Gabriel Alfonso Rincón-Mora

School of Electrical and Computer Engineering
Georgia Institute of Technology

Rincon-Mora.gatech.edu

Copyright © 2019 by G.A. Rincón-Mora

Copyright © 2019 by Gabriel A. Rincón-Mora. All rights reserved.

No part of this publication may be reproduced, stored in a retrieval system or transmitted in any form or by any means, electronic, mechanical, photocopying, recording, scanning or otherwise, except as permitted under Section 107 or 108 of the 1976 United States Copyright Act, without the prior written permission of the author.

Contents

	Page
List of Figures	v
List of Abbreviations	vii
1. Junction FETs	1
1.1. N Channel	1
A. Triode	2
B. Saturation	3
C. I–V Translation	4
D. Symbol	5
1.2. P Channel	5
A. Triode	6
B. Saturation	7
C. I–V Translation	7
D. Symbol	8
2. N-Channel MOSFETs	8
2.1. Accumulation: Cut Off	9
2.2. Depletion: Sub-Threshold	9
A. Triode	11
B. Saturation	12
C. I–V Translation	12
2.3. Inversion	13
A. Triode	13
B. Saturation	15
C. I–V Translation	16
2.4. Body Effect	17
2.5. Symbols	19
3. P-Channel MOSFETs	19
3.1. Accumulation: Cut Off	20
3.2. Depletion: Sub-Threshold	20
A. Triode	22
B. Saturation	23
C. I–V Translation	23
3.3. Inversion	23
A. Triode	23
B. Saturation	25
C. I–V Translation	27
3.4. Body Effect	28
3.5. Symbols	28
3.6. Unifying Convention	29
4. Capacitances	30

4.1. PN-Junction Capacitances	30
4.2. Gate-Oxide Capacitances	30
A. Cut Off	31
B. Sub-Threshold	31
C. Triode Inversion	32
D. Saturated Inversion	32
E. Transition	33
4.3. MOS Varactors	34
A. Bi-Modal	34
B. Inversion Mode	35
C. Accumulation Mode	35
D. Variations	36
4.4. MOS Diodes	36
A. Diode Connection	36
B. Diode Action	38
5. Short Channels	**39**
5.1. Drain-Induced Barrier Lowering	39
A. Thinner Oxide	41
5.2. Gate–Channel Field	41
A. Surface Scattering	41
B. Hot-Electron Injection	42
C. Oxide-Surface Ejections	42
D. Fringing Fields	42
5.3. Source–Drain Field	43
A. Velocity Saturation	43
B. Impact Ionization	44
C. Arcing Field	45
D. Lightly-Doped Drain	46
6. Other Considerations	**47**
6.1. Weak Inversion	47
A. Voltage Bias	47
B. Current Bias	48
6.2. Junction Isolation	51
A. Channel BJTs	52
B. Substrate BJTs	53
C. Substrate MOSFETs	53
D. Welled MOSFETs	54
E. Process Variants	54
6.3. Diffused-Channel MOSFETs	54
6.4. Noise	56
A. Terminology	56
B. Electronic Noise	57
C. Systemic Noise	59
7. Summary	**60**

List of Figures

	Page
Figure 1. N-channel JFET structure.	2
Figure 2. Uniformly biased N-channel JFET in triode.	2
Figure 3. Asymmetrically biased N-channel JFET in triode.	3
Figure 4. N-channel JFET in saturation.	3
Figure 5. N-channel JFET current.	5
Figure 6. N-channel JFET symbol.	5
Figure 7. P-channel JFET structure.	6
Figure 8. P-channel JFET in triode.	6
Figure 9. P-channel JFET current.	7
Figure 10. P-channel JFET symbol.	8
Figure 11. N-channel MOSFET structure.	9
Figure 12. N-channel MOSFET in accumulation and cut off.	9
Figure 13. N-channel MOSFET in depletion.	9
Figure 14. Band diagram of N-channel MOSFET in depletion.	10
Figure 15. Band diagram of N-channel MOSFET in sub-threshold.	10
Figure 16. Voltage divider across gate oxide and surface–body.	11
Figure 17. Sub-threshold N-channel MOSFET current.	13
Figure 18. Symmetrically inverted N-channel MOSFET in triode.	13
Figure 19. Asymmetrically inverted N-channel MOSFET in triode.	15
Figure 20. Inverted N-channel MOSFET in saturation.	15
Figure 21. Inverted N-channel MOSFET current.	16
Figure 22. N-channel MOSFET symbols.	19
Figure 23. P-channel MOSFET structure.	20
Figure 24. P-channel MOSFET in accumulation and cut off.	20
Figure 25. P-channel MOSFET in depletion.	21
Figure 26. Band diagram of P-channel MOSFET in depletion.	21
Figure 27. Band diagram of P-channel MOSFET in sub-threshold.	21
Figure 28. Sub-threshold P-channel MOSFET current.	23
Figure 29. Symmetrically inverted P-channel MOSFET in triode.	24
Figure 30. Asymmetrically inverted P-channel MOSFET in triode.	25
Figure 31. Inverted P-channel MOSFET in saturation.	26
Figure 32. Inverted P-channel MOSFET current.	27
Figure 33. P-channel MOSFET symbols.	29
Figure 34. Gate-oxide capacitances.	31

Figure 35. Inverted gate–source and gate–drain capacitances.   33
Figure 36. Bi-modal P-channel MOSFET varactor.   34
Figure 37. Inversion-mode P-channel MOSFET varactor.   35
Figure 38. Accumulation-mode N-channel MOSFET varactor.   35
Figure 39. N- and P-channel MOSFET diodes.   37
Figure 40. Implicit N- and P-channel MOS diode action.   38
Figure 41. Drain-induced punch-through.   40
Figure 42. Drain-induced barrier lowering in an N-channel MOSFET.   40
Figure 43. Channel coupling components.   41
Figure 44. Surface scattering and hot-electron injection in the NMOS.   41
Figure 45. Fringing electrics-field lines around an N-channel MOSFET.   43
Figure 46. Velocity saturation and pinch-off effects in inversion.   44
Figure 47. Impact ionization, avalanche, and hot-electron injection.   45
Figure 48. Lightly-doped-drain MOSFETs   46
Figure 49. Drain current as MOSFET channel forms.   47
Figure 50. Small-signal transconductance across regions.   50
Figure 51. Junction-isolated single-well P-substrate CMOS FETs.   52
Figure 52. Substrate NMOS with parasitic components.   53
Figure 53. Welled PMOS with parasitic components.   54
Figure 54. Lateral diffused-channel N-channel MOSFET.   55
Figure 55. Spectrum of flicker and thermal noise.   58

List of Abbreviations

BJT ≡ Bipolar-Junction Transistor
CMOS ≡ Complementary MOS
DIBL ≡ Drain-Induced Barrier Lowering
DMOS ≡ Diffused-Channel or Double-Diffused MOS
EHP ≡ Electron–Hole Pair
FET ≡ Field-Effect Transistor
JFET ≡ Junction FET
LDMOS ≡ Lateral DMOS
LDD ≡ Lightly Doped Drain
MOS ≡ Metal–Oxide–Semiconductor
NBTI ≡ Negative Bias Temperature Instability
RSS ≡ Root Sum of Squares
SNR ≡ Signal-to-Noise Ratio
SiO_2 ≡ Silicon Dioxide
VCO ≡ Voltage-Controlled Oscillator
VDMOS ≡ Vertical DMOS

A_J ≡ Junction Area

β_0 ≡ Base–Collector Current Gain

C_{CH} ≡ Channel Capacitance
C_{DEP} ≡ Channel–Body Depletion Capacitance
C_J ≡ Junction Capacitance
C_{J0} ≡ Zero-Bias C_J
C_{OL} ≡ Overlap Capacitance
C_{OX} ≡ Oxide Capacitance

d_W ≡ Depletion Width

E_{BG} ≡ Band-Gap Energy
$E_{CN/P}$ ≡ Critical Electric Field
E_F ≡ Fermi Energy Level
E_K ≡ Kinetic Energy
ε_0 ≡ Permittivity in Vacuum
ε_{OX} ≡ Relative Permittivity of SiO_2

f_C ≡ Noise Corner Frequency
f_O ≡ Operating Frequency
f_{SW} ≡ Switching Frequency
Δf_{BW} ≡ Frequency Bandwidth

MOSFETs

$g_m \equiv$ Small-Signal Transconductance
$\gamma_{N/P} \equiv$ Body-Effect Parameter

$i_B \equiv$ Body Current
$i_D \equiv$ Drain Current
$i_{DIF} \equiv$ Diffusion Current
$i_{FLD} \equiv$ Drift Current
$i_G \equiv$ Gate Current
$i_{IN} \equiv$ Input Current
$i_{nc} \equiv$ Coupled Noise Current
$i_{nf} \equiv$ Flicker Noise Current
$i_{ns} \equiv$ Shot Noise Current
$i_{nt} \equiv$ Thermal Noise Current
$i_S \equiv$ Source Current
$i_{SUB} \equiv$ Substrate Current
$I_{SN/P} \equiv$ Saturation Current

$K_B \equiv$ Boltzmann's Constant
$K_F \equiv$ Flicker Noise Coefficient
$K_{N/P} \equiv$ Baseline Conductivity
$K_{N/P}' \equiv$ Transconductance Parameter

$L_{CH} \equiv$ Channel Length
$L_{MIN} \equiv$ Minimum Oxide Length
$L_{OL} \equiv$ Overlap Length
L_{OX} or $L \equiv$ Oxide Length
$\lambda_{N/P} \equiv$ Channel-Length Modulation Parameter

$\mu_N \equiv$ Electron Mobility
$\mu_P \equiv$ Hole Mobility

$n_d \equiv$ Spectral Noise Density
$n_I \equiv$ Non-Ideality Factor

$q_E \equiv$ Electronic Charge

$R_{CH} \equiv$ Channel Resistance
$R_{ON} \equiv$ On Resistance
$R_{SD} \equiv$ Source–Drain Resistance
$R_{SH} \equiv$ Sheet Resistivity

$t_{OX} \equiv$ Oxide Thickness
$T_J \equiv$ Junction Temperature

$v_B \equiv$ Body Terminal/Voltage
$v_{BS} \equiv$ Body–Source Voltage

MOSFETs

$v_D \equiv$ Drain Terminal/Voltage
$v_{DD} \equiv$ Positive Power Supply
$v_{DS} \equiv$ Drain–Source Voltage
$v_{DS(SAT)} \equiv$ Drain–Source Pinch-Off Saturation Voltage
$v_{DS(SAT)}' \equiv$ Drain–Source Sub-Threshold Saturation Voltage
$v_{DS(SAT)}'' \equiv$ Drain–Source Velocity-Saturation Voltage
$v_E \equiv$ Electron Velocity
$v_G \equiv$ Gate Terminal/Voltage
$v_{GS} \equiv$ Gate–Source Voltage
v_{GSP}, v_{GST}, and $v_{SGT} \equiv$ Gate Drive
$v_H \equiv$ Hole Velocity
$v_{JR} \equiv$ Reverse Junction Voltage
$v_{nt} \equiv$ Thermal Noise Voltage
$v_S \equiv$ Source Terminal/Voltage
$v_{TN/P} \equiv$ Threshold Voltage
$V_A \equiv$ Early Voltage
$V_{BI} \equiv$ Built-In Potential
$V_P \equiv$ Pinch-Off Voltage
$V_t \equiv$ Thermal Voltage
$V_{TN/P0} \equiv$ Zero-Bias Threshold
$\psi_B \equiv$ Surface–Body Barrier
$\psi_S \equiv$ Surface Potential

$w_B \equiv$ Effective Base Width
W_{CH} or $W \equiv$ Channel Width

MOSFETs

MOSFETs

Power supplies use switches to steer and feed current into batteries and microelectronic systems. *Metal–oxide–semiconductor* (MOS) *field-effect transistors* (FETs) are popular in this space because they drop millivolts and do not require static gate current to close. Although *bipolar-junction transistors* (BJTs) usually cost less to fabricate, they need static base current to close and saturate when they close, so they require substantial reverse current to open. Diodes do not need this static current, but they close conditionally and drop 400–700 mV. Still, MOSFETs incorporate substrate diodes and BJTs that can help, and at times also hurt.

The fundamental mechanism that establishes conductivity in FETs is an electric field. In the case of MOSFETs, parallel-plate MOS capacitors establish this field. The underlying purpose of the capacitor is to form a conducting *N-type channel* in NFETs and a *P-type channel* in PFETs. Although poly-silicon is nowadays more popular than metal as the upper plate, engineers still use MOS to refer to these and other oxide-sandwiched structures on semiconductors.

1. Junction FETs

The *junction FET* (JFET) is a simpler junction-based realization of the MOSFET. Although not as pervasive, JFETs are useful as resistors and low-noise transistors. *Electronic noise* in JFETs is low because carriers flow well below (and away from) the uneven silicon surface.

1.1. N Channel

N-channel JFETs are N-doped semiconductor strips sandwiched between P-type regions. In the case of Fig. 1, *top* and *bottom* P^+ and P *gates* sandwich an N channel contacted by highly doped N^+ regions. *Channel length* L$_{CH}$ or L and *width* W$_{CH}$ or W are the longitudinal length and width of the overlapping P$^+$–N channel–P gate layers. The Ohmic surface contact of the bottom gate is another highly doped P$^+$ region.

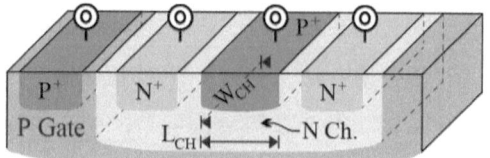

Fig. 1. N-channel JFET structure.

A. Triode

The NJFET is basically an N resistor compressed by P regions. The geometry and doping concentration of the channel set baseline *channel resistance* R$_{CH}$. R$_{CH}$ climbs as the depletion space against the top and bottom P regions in Fig. 2 expand to squeeze the channel. These P regions are the *gates* v$_G$ of the JFET because their voltage adjusts R$_{CH}$.

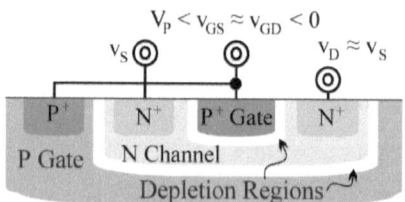

Fig. 2. Uniformly biased N-channel JFET in triode.

R$_{CH}$ is high when L is long, W is narrow, and *baseline conductivity* K$_N$ is low. The channel dematerializes (and opens) when the gate–channel voltage reverses enough to pinch the entire channel. This negative v$_G$ is the *pinch-off voltage* V$_P$. So R$_{CH}$ spikes sharply when v$_{GS}$ and v$_{GD}$ reach this V$_P$:

$$R_{CH}\bigg|_{v_{DS}\ll v_{GSP}}^{v_{GS}>V_P} \approx \left(\frac{L}{W}\right)\left[\frac{1}{K_N(v_{GS}-V_P)}\right] \equiv \left(\frac{L}{W}\right)\left(\frac{1}{K_N v_{GSP}}\right) \equiv \left(\frac{L}{W}\right) R_{SH}. \quad (1)$$

This R$_{CH}$ is more accurate when v$_{GS}$ is uniform across the channel and above the pinch-off point. This happens when v$_{GS}$ and v$_{GD}$ or v$_{GS}$ –

v_{DS} are higher than V_P and v_{DS} is low. This means that *gate drive* $v_{GS} - V_P$ or v_{GSP} is positive and v_{DS} is much lower than v_{GSP}.

K_N and v_{GSP} set the *sheet resistivity* R_{SH} of R_{CH} (in Ohms per square). R_{SH} is the resistance of each W × W square of the channel. So when combined, R_{CH} is L/W squares of R_{SH}.

A voltage v_{DS} across R_{CH} induces channel current i_D. Except, v_{DS} expands the depletion space that squeezes half of the channel in Fig. 3. So half v_{DS} raises and reduces the linear effects of v_{GS} on R_{CH} and i_D:

$$i_D \Big|_{v_{DS}<v_{GSP}}^{v_{GS}>V_P} = \frac{V_{DS}}{R_{CH}} \approx v_{DS}\left(\frac{W}{L}\right)K_N\left(v_{GS} - V_P - \frac{v_{DS}}{2}\right). \quad (2)$$

This i_D corresponds to *triode* because i_D is sensitive to both v_{SG} and v_{SD}.

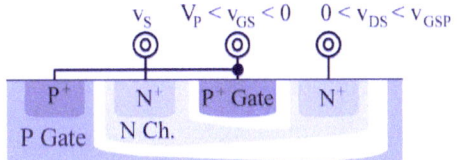

Fig. 3. Asymmetrically biased N-channel JFET in triode.

With v_{DS}, the negative N^+ terminal supplies the N-channel electrons that the positive N^+ terminal outputs. So the N^+ region on the left is the *source* v_S and the one on the right is the *drain* v_D. But when v_G connects to v_S or a fixed negative voltage, the NJFET is just a *pinched resistor*.

B. Saturation

The channel pinches and charge disappears near v_D in Fig. 4 when v_{GD} reverses to V_P. Since v_{GD} is equivalent to $v_{GS} - v_{DS}$, this happens when v_{DS} reaches $v_{GS} - V_P$ or v_{GSP}. This is the *saturation voltage* $v_{DS(SAT)}$.

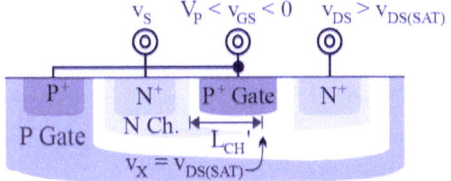

Fig. 4. N-channel JFET in saturation.

When v_{DS} overcomes this $v_{DS(SAT)}$, the channel extends to the fraction of L that drops $v_{DS(SAT)}$: L_{CH}'. The depleted portion of L drops what remains of v_{DS}: $v_{DS} - v_{DS(SAT)}$. i_D saturates because the voltage across the channel that L_{CH}' establishes is $v_{DS(SAT)}$, which is independent of v_{DS}:

$$i_D\Big|_{\substack{v_{GS} > V_P \\ v_{DS} > v_{GSP}}} = \frac{V_{DS(SAT)}}{R_{CH}} \approx v_{DS(SAT)} \left(\frac{W}{L_{CH}'}\right) K_N \left(v_{GS} - V_P - \frac{V_{DS(SAT)}}{2}\right)$$

$$\approx \left(\frac{W}{L}\right)\left(\frac{K_N}{2}\right)(v_{GS} - V_P)^2 (1 + \lambda_N v_{DS}). \quad (3)$$

This i_D corresponds to *saturation* because i_D is sensitive to v_{GS} and largely insensitive to v_{DS}. In other words, i_D saturates with respect to v_{DS}.

Since a higher v_D depletes more of the channel, L_{CH}' shrinks as v_{DS} increases, which means R_{CH} falls and i_D increases. This is *channel-length modulation*. Channel-length modulation parameter λ_N models this effect with respect to the L that circuit designers define.

Interestingly, the effects of v_{DS} fade as L lengthens. This is because the variation in L_{CH}' becomes a smaller fraction of L. So *drain current* i_D becomes less sensitive to v_{DS}, which is another way of saying λ_N is lower.

C. I–V Translation

i_D is sensitive to v_{DS} in triode and insensitive to v_{DS} in saturation like Fig. 5 shows. Since N^+ regions can reverse roles, negative v_{GD} and v_{DS} establish a negative i_D that mirrors the i_D that a negative v_{GS} and a positive v_{DS} produce. i_D is zero (in *cut off*) when v_{GS} and v_{GD} reach V_P. The x-axis represents this mode because i_D is zero along that line.

i_D saturates when v_{DS} overcomes $v_{DS(SAT)}$'s v_{GSP}. Since i_D is a quadratic translation of v_{GSP} in saturation, the $v_{DS(SAT)}$ boundary in Fig. 5 is a squared-root reflection of i_D:

$$V_{DS(SAT)} = v_{GS} - V_P \approx \sqrt{\frac{2i_D}{(W/L)K_N(1 + \lambda_N v_{DS})}}. \quad (4)$$

$\lambda_{NV_{DS}}$ fades with longer L's because L_{CH}' modulation diminishes.

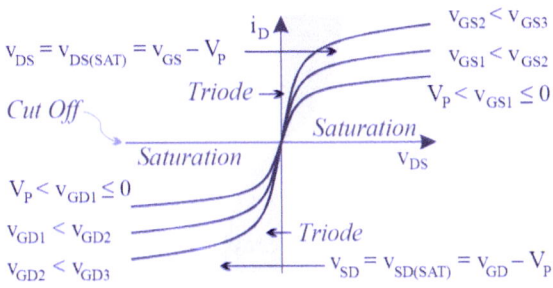

Fig. 5. N-channel JFET current.

D. Symbol

JFETs are three-terminal devices with interchangeable v_S and v_D terminals that conduct i_D in Fig. 6 when v_{GS} is zero or negative and v_{DS} is positive. The two vertical lines (into which v_G connects) symbolize the PN depletion capacitance that pinches the channel. *Gate current* i_G is close to zero because v_G reverse-biases the gate–channel junction. So *source current* i_S outputs almost all the i_D that v_D receives. The arrow indicates the P gate and N channel form a PN junction.

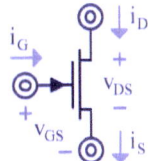

Fig. 6. N-channel JFET symbol.

1.2. P Channel

PJFETs are and operate exactly the same way as NJFETs, except with a P channel. So top and bottom N^+ and N gates in Fig. 7 sandwich a P channel contacted by highly doped P^+ regions. L_{CH} or L and W_{CH} or W are the longitudinal length and width of the overlapping N^+–P Channel–N gate layers. A highly doped N^+ region contacts the bottom gate.

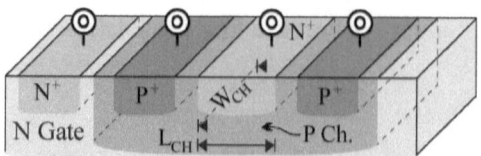

Fig. 7. P-channel JFET structure.

A. Triode

The PJFET is basically a P resistor compressed by N regions. R_{CH} is high when L is long, W is narrow, and K_P is low. The channel dematerializes (opens) when the gate–channel voltage reverses enough to pinch the entire channel. So R_{CH} spikes sharply when v_{SG} and v_{DG} reach V_P:

$$R_{CH}\bigg|_{v_{SD}\ll v_{SGP}}^{v_{SG}>V_P} \approx \left(\frac{L}{W}\right)\left[\frac{1}{K_P(v_{SG}-V_P)}\right] \equiv \left(\frac{L}{W}\right)\left(\frac{1}{K_P v_{SGP}}\right) \equiv \left(\frac{L}{W}\right) R_{SH}. \quad (5)$$

R_{CH} is more accurate when v_{SG} is uniform across the channel and above the pinch-off point. This happens when v_{SG} and v_{DG} or $v_{SG} - v_{SD}$ are higher than V_P and v_{SD} is low. This means that $v_{SG} - V_P$ or v_{SGP} is positive and v_{SD} is much lower than v_{SGP}.

A voltage v_{SD} across R_{CH} induces i_D. Except, v_{SD} expands the depletion space that squeezes half of the channel in Fig. 8. So half v_{DS} raises and reduces the linear effects of v_{SG} on R_{CH} and i_D:

$$i_D\bigg|_{v_{SD}<v_{SGP}}^{v_{SG}>V_P} = \frac{V_{SD}}{R_{CH}} \approx v_{SD}\left(\frac{W}{L}\right)K_P\left(v_{SG}-V_P-\frac{V_{SD}}{2}\right). \quad (6)$$

Since the positive P^+ supplies the P-channel holes that the negative P^+ outputs, the P^+ on the left is v_S and the one on the right is v_D.

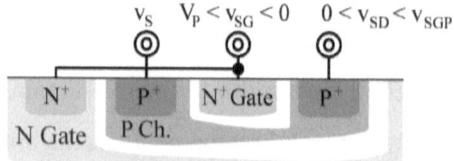

Fig. 8. P-channel JFET in triode.

B. Saturation

The channel pinches and charge disappears near v_D when v_{DG} reverses to V_P. Since v_{DG} is equivalent to $v_{SG} - v_{SD}$, this happens when v_{SD} reaches the $v_{SD(SAT)}$ that $v_{SG} - V_P$ or v_{SGP} set. When v_{SD} surpasses this $v_{SD(SAT)}$, the channel extends to the fraction of L that drops $v_{SD(SAT)}$: L_{CH}'. The depleted portion of L drops the remainder: $v_{SD} - v_{SD(SAT)}$. So i_D saturates to the level that $v_{SD(SAT)}$ across L_{CH}' sets:

$$i_D \Big|_{\substack{v_{SG} > V_P \\ v_{SD} > v_{SGP}}} = \frac{v_{SD(SAT)}}{R_{CH}} \approx v_{SD(SAT)} \left(\frac{W}{L_{CH}'}\right) K_P \left(v_{SG} - V_P - \frac{v_{SD(SAT)}}{2}\right)$$
$$\approx \left(\frac{W}{L}\right)\left(\frac{K_P}{2}\right)(v_{SG} - V_P)^2 (1 + \lambda_P v_{SD}) \quad (7)$$

L_{CH}' shrinks as v_{SD} climbs because a lower v_D depletes more of the channel. This means that R_{CH} falls and i_D increases. Lengthening L reduces λ_P because variations in L_{CH}' become a smaller fraction of L.

C. I–V Translation

i_D is sensitive to v_{SD} in triode and insensitive to v_{SD} in saturation like Fig. 9 shows. Since P$^+$ regions can reverse roles, negative v_{DG} and v_{SD} establish a negative i_D that mirrors the i_D that a negative v_{SG} and a positive v_{SD} produce. i_D is zero when v_{SG} and v_{DG} reach V_P.

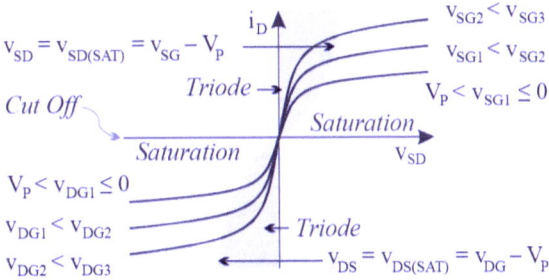

Fig. 9. P-channel JFET current.

i_D saturates when v_{SD} overcomes $v_{SD(SAT)}$'s v_{SGP}. Since i_D is a quadratic translation of v_{SGP} in saturation, the $v_{SD(SAT)}$ boundary in Fig. 9 is a squared-root reflection of i_D:

$$V_{SD(SAT)} = V_{SG} - V_P \approx \sqrt{\frac{2i_D}{(W/L)K_P(1+\lambda_P V_{SD})}}. \tag{8}$$

$\lambda_P v_{SD}$ fades with longer L's because L_{CH}' modulation diminishes.

D. Symbol

PJFETs conduct i_D in Fig. 10 when v_{SG} is zero or negative and v_{SD} is positive. i_G is close to zero because v_G reverse-biases the gate–channel junction. So i_D outputs nearly all the i_S that v_S receives. The arrow indicates the N gate and P channel form a PN junction and the parallel lines next to it symbolize the junction capacitance that results.

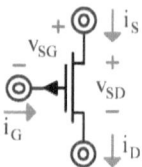

Fig. 10. P-channel JFET symbol.

2. N-Channel MOSFETs

The N-channel MOSFET is a smaller and slightly more involved NJFET. Structurally, a MOS capacitor is on P material that constitutes the *body* or *bulk* of the transistor. The *thin oxide* SiO_2 in the MOS capacitor in Fig. 11 sets the *oxide length* L_{OX} or L of the FET, and together with the N^+ regions, the width of the channel W_{CH} or W to be established under this oxide. L_{CH} is the distance between these highly doped N^+ regions. The N^+ terminals extend into the oxide region (across *overlap length* L_{OL}) to ensure they connect to the N channel that the oxide field forms. The highly doped P^+ region is an Ohmic surface contact to the body.

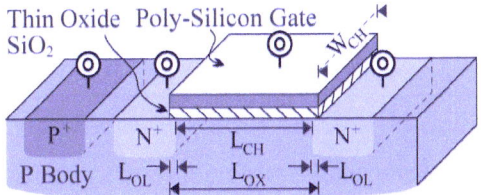

Fig. 11. N-channel MOSFET structure.

2.1. Accumulation: Cut Off

The P body usually connects to the most negative potential (ground in Fig. 12) to keep body–N^+ junctions from forward biasing. So the electrons and holes that diffuse recombine and deplete the regions near those junctions. Applying a negative v_G to the poly-silicon gate pulls holes in the body toward the oxide. Holes therefore accumulate near the surface of the semiconductor. This way, in *accumulation*, current does not flow, so the NFET is in cut off.

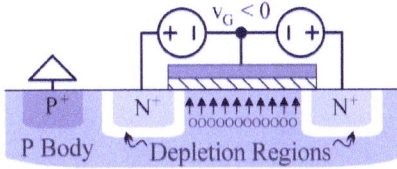

Fig. 12. N-channel MOSFET in accumulation and cut off.

2.2. Depletion: Sub-Threshold

Applying a positive v_G does the opposite: pushes holes away from the semiconductor surface in Fig. 13, which depletes the region under the oxide of holes. With fewer holes with which to recombine, N^+ electrons diffuse farther before recombining. A positive v_G also pulls and loosens electrons from their N^+ home sites. So v_G reduces the barrier voltage that keeps N^+ electrons from diffusing and extends their diffusion length.

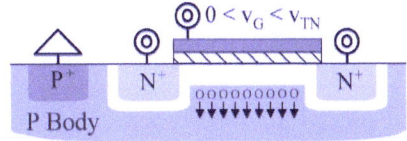

Fig. 13. N-channel MOSFET in depletion.

In depletion, the carrier density near the surface under the oxide falls to the point of becoming nearly intrinsic. This is why the *Fermi level* E_F in this region in Fig. 14 is nearly halfway across the *band gap* E_{BG}. But without any voltage between the N^+ regions, current does not flow.

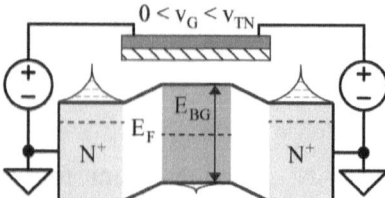

Fig. 14. Band diagram of N-channel MOSFET in depletion.

Raising the voltage of one of the N^+ terminals (v_D in Fig. 15) elevates the barrier and expands the depletion region around that terminal. The resulting electric field E_{FLD} pulls diffusing electrons into v_D. As E_{FLD} intensifies, more electrons (that would otherwise recombine) reach v_D. Recombination nearly stops when v_D is 3 to 4 *thermal voltages* V_t's above v_S, which is equivalent to 75 to 100 mV or so at room temperature.

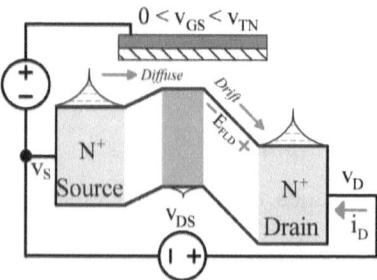

Fig. 15. Band diagram of N-channel MOSFET in sub-threshold.

v_D is the drain because the N^+ terminal with the higher potential drains these diffusing electrons. And v_S is the source because the terminal with the lower potential supplies these electrons. The resulting flow of electrons establishes an i_D that flows into v_D and out of v_S.

A. Triode

i_D rises with v_{GS} because a positive gate–source voltage reduces the gate–source barrier. This rise is exponential because v_{GS} avails exponentially more electrons than *junction temperature* T_J avails with V_t:

$$i_D\big|^{0<v_{GS}<v_{TN}} = \left(\frac{W}{L}\right) I_{SN} \exp\left[\left(\frac{C_{OX}"}{C_{OX}"+C_{DEP}"}\right)\left(\frac{V_{GST}}{V_t}\right)\right]\left[1-\exp\left(\frac{-V_{DS}}{V_t}\right)\right]$$

$$= \left(\frac{W}{L}\right) C_{DEP}" \mu_N V_t^2 \exp\left(\frac{V_{GS}-V_{TN}}{n_I V_t}\right)\left[1-\frac{1}{\exp(v_{DS}/V_t)}\right]. \quad (9)$$

i_D also increases with v_{DS} because v_{DS} intensifies the field that pulls them to v_D. This i_D corresponds to *triode* because i_D is sensitive to v_{GS} and v_{DS}.

i_D is high when W is wide, L is short, and the charge held in the *channel–body depletion capacitance* (per unit area) $C_{DEP}"$ is high. These, *electron mobility* μ_N, and V_t set the baseline conductivity of i_D. i_D's *saturation current* I_{SN} combines the effects of $C_{DEP}"$, μ_N, and V_t.

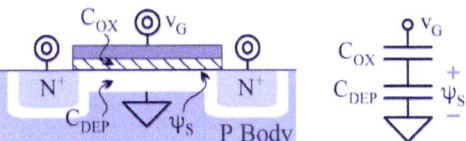

Fig. 16. Voltage divider across gate oxide and surface–body.

The surface-to-body *surface potential* ψ_S in Fig. 16 is ultimately the voltage-divided fraction that v_G couples through the *oxide capacitance* (per unit area) $C_{OX}"$ into $C_{DEP}"$, where

$$C_{OX}" = \frac{\varepsilon_{OX}}{t_{OX}} = \frac{\varepsilon_{Si}\varepsilon_0}{t_{OX}} = \frac{3.9\varepsilon_0}{t_{OX}}, \quad (10)$$

permittivity in vacuum ε_0 is 8.845 pF/m, *permittivity of silicon dioxide* ε_{OX} is the *relative permittivity of silicon* ε_{Si} times ε_0 or $3.9\varepsilon_0$, and *oxide thickness* t_{OX} is on the order of nanometers. The *non-ideality factor* n_I in i_D models the reduction in gate drive v_{GST} or $v_{GS} - v_{TN}$ that $C_{OX}"$ and

C_{DEP}" cause. Surface imperfections also impede diffusion, so in practice, n_I also accounts for these defects. n_I is typically between 1 and 2.

Native or *natural* NFETs conduct some i_D when v_{GS} is zero. Normally-on devices like these are *depletion-mode transistors*. They are useful in low-voltage applications that benefit from low v_{GS} values.

Many analog and almost all digital circuits, however, require an off state. So process engineers oftentimes implant additional acceptor dopant atoms below the oxide. With more holes, conduction requires a higher v_{GS}. This implant therefore reduces the exponential effect of v_{GS}, which raises the *threshold voltage* v_{TN} to, in practice, 400–600 mV. Devices that can turn on and off are *enhancement-mode transistors*.

B. Saturation

i_D climbs with v_{DS} because v_{DS} intensifies the field that reduces recombination. Diffusion length is so high in this quasi-intrinsic region that nearly 95% of the electrons reach the other end when v_{DS} is $3V_t$. Past this point, i_D saturates to

$$i_D \Big|_{v_{DS}>3V_t}^{0<v_{GS}<v_{TN}} \approx \left(\frac{W}{L}\right) I_{SN} \exp\left(\frac{v_{GS}-v_{TN}}{n_I V_t}\right), \tag{11}$$

where $v_{DS(SAT)}$' is the *sub-threshold saturation voltage* that $3V_t$ sets. This i_D refers to the saturation region because i_D is insensitive to v_{DS}.

C. I–V Translation

i_D in Fig. 17 is sensitive to v_{DS} in triode and insensitive to v_{DS} in saturation, when v_{DS} is greater than $3V_t$ or so. N^+ regions can change roles and conduct reverse current because the structure is symmetrical. So a positive v_{GD} and a negative v_{DS} establish a negative i_D that mirrors the i_D that positive v_{GS} and v_{DS} produce. i_D is zero in accumulation, when v_{GS} and v_{GD} are both negative.

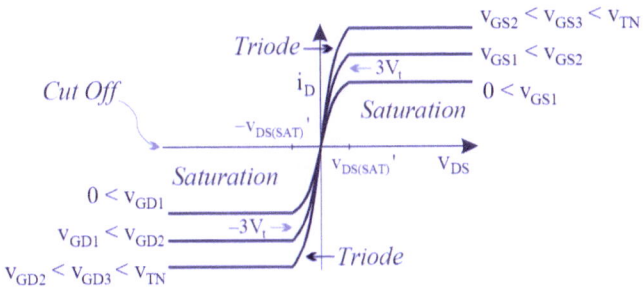

Fig. 17. Sub-threshold N-channel MOSFET current.

2.3. Inversion

A. Triode

Raising v_{GS} and v_{GD} first eliminates the barrier that keeps electrons from flowing. Past this barrier point, the electric field across the oxide is high enough to pull electrons from both N⁺ terminals in Fig. 18 toward the region below the gate. When v_{GS} and v_{GD} reach v_{TN}, these electrons establish a conducting N channel with a carrier density that matches that of the original zero-bias P body. So overcoming this *threshold voltage* v_{TN} inverts this medium. This is the process to which *inversion* refers.

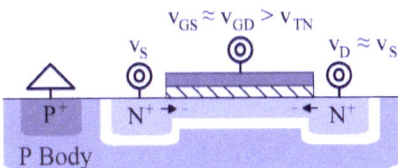

Fig. 18. Symmetrically inverted N-channel MOSFET in triode.

R_{CH} is high when L is long, W is narrow, and the conductivity that μ_N and C_{OX}'' in *transconductance parameter* K_N' set is low. The channel dematerializes when v_{GS} and v_{GD} drop below v_{TN}. So R_{CH} spikes sharply when v_{GS} and v_{GD} reach v_{TN}:

$$R_{CH}\Big|_{v_{DS} \ll v_{GST}}^{v_{GS} > v_{TN}} \approx \left(\frac{L}{W}\right)\left[\frac{1}{\mu_N C_{OX}''(v_{GS} - v_{TN})}\right]$$

$$\equiv \left(\frac{L}{W}\right)\left(\frac{1}{K_N'v_{GST}}\right) \equiv \left(\frac{L}{W}\right)R_{SH}. \quad (12)$$

R_{CH} is more accurate when v_{GS} and v_{GD} match and overcome v_{TN}. So v_{GS} and v_{GD} or $v_{GS} - v_{DS}$ are higher than v_{TN} and v_{DS} is low. This means that gate drive $v_{GS} - v_{TN}$ or v_{GST} is positive and v_{DS} is much lower than v_{GST}. K_N' and v_{GST} set the sheet resistivity R_{SH} of R_{CH}.

Example 1: Determine R_{CH} when W is 10 µm, L is 180 nm, v_{GS} is 1.8 V, v_{DS} is 50 mV, μ_N is 72k mm²/V·s, t_{OX} is 12.5 nm, L_{OL} is 30 nm, and v_{TN} is 400 mV.

Solution:

$$C_{OX}'' = \frac{\varepsilon_{OX}}{t_{OX}} = \frac{3.9\varepsilon_0}{t_{OX}} = \frac{3.9(8.845 \times 10^{-12})}{12.5n} = 2.76 \text{ fF}/\mu m^2$$

$$K_N' = C_{OX}''\mu_N = (2.76m)(72m) = 200 \text{ }\mu A/V^2$$

$$v_{DS} = 50 \text{ mV} \ll v_{GS} - v_{TN} = 1.8 - 400m = 1.4 \text{ V} \therefore$$

$$R_{CH} \approx \left(\frac{L - 2L_{OL}}{W_{CH}}\right)\left[\frac{1}{K_N'(v_{GS} - v_{TN})}\right]$$

$$= \left[\frac{180n - 2(30n)}{10\mu}\right]\left[\frac{1}{(200\mu)(1.8 - 400m)}\right] = 43 \text{ }\Omega$$

---

v_{DS} across this R_{CH} induces i_D. But since v_D opposes the inversion action of v_{GD} across half the channel, half of v_{DS} in Fig. 19 opposes the linear effects of v_{GS}. So R_{CH} rises and i_D falls with $0.5v_{DS}$:

$$i_D\Big|_{v_{DS}<v_{GST}}^{v_{GS}>v_{TN}} = \frac{v_{DS}}{R_{CH}} = v_{DS}\left(\frac{W}{L}\right)K_N'\left(v_{GS} - v_{TN} - \frac{v_{DS}}{2}\right). \quad (13)$$

i_D climbs almost linearly with v_{GS} because an electric field (not diffusion) alters conductivity. This i_D corresponds to triode inversion because i_D is sensitive to v_{GS} and v_{DS}.

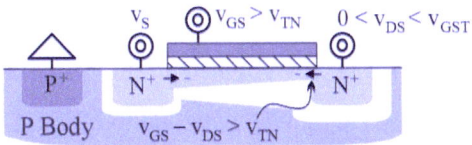

Fig. 19. Asymmetrically inverted N-channel MOSFET in triode.

B. Saturation

Charge q_D near the edge of v_D's N$^+$ region in Fig. 20 disappears when v_{GD} falls to v_{TN}. Since v_{GD} is v_{TN} when this happens and v_{GD} is equivalent to $v_{GS} - v_{DS}$, the channel pinches when v_{DS} reaches $v_{GS} - v_{TN}$. So gate drive sets the saturation voltage $v_{DS(SAT)}$ of the inverted NFET.

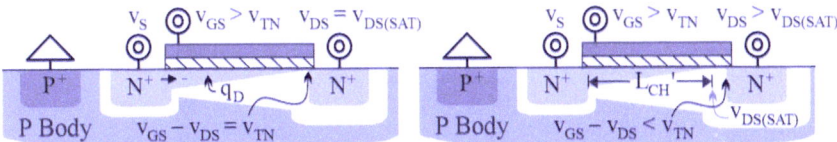

Fig. 20. Inverted N-channel MOSFET in saturation.

When v_{DS} overcomes $v_{DS(SAT)}$, the channel extends to the fraction of L_{CH} that drops $v_{DS(SAT)}$: L_{CH}'. The depleted portion of L_{CH} drops what remains of v_{DS}: $v_{DS} - v_{DS(SAT)}$. i_D saturates because the voltage across the channel that L_{CH}' establishes is $v_{DS(SAT)}$, which is independent of v_{DS}:

$$i_D\Big|_{\substack{v_{GS}>v_{TN}\\v_{DS}>v_{GST}}} = \frac{V_{DS(SAT)}}{R_{CH}} = v_{DS(SAT)} \left(\frac{W}{L_{CH}'}\right) K_N' \left(v_{GS} - v_{TN} - \frac{v_{DS(SAT)}}{2}\right)$$

$$\approx \left(\frac{W}{L}\right)\left(\frac{K_N'}{2}\right)(v_{GS} - v_{TN})^2 (1 + \lambda_N v_{DS}) \qquad . \quad (14)$$

$$\equiv \left(\frac{W}{L}\right)\left(\frac{K_N'}{2}\right) v_{GST}^2 \left(1 + \frac{v_{DS}}{V_{AN}}\right)$$

In other words, i_D is sensitive to v_{GS} and largely insensitive to v_{DS}.

Since a higher v_D depletes more of the channel, L_{CH}' shrinks as v_{DS} rises. So R_{CH} falls and i_D increases with v_{DS} according to λ_N with respect to the L that circuit designers define. v_{DS} effects fade (and λ_N falls) as L lengthens because the variation in L_{CH}' becomes a smaller fraction of L.

Although different in nature, the effect of λ_N mirrors *base-width modulation* in BJTs. This is why engineers often use *Early voltage* V_A to quantify this effect in FETs. So when applied to NFETs, V_{AN} is $1/\lambda_N$.

Since C_{GS} falls with L_{CH}', v_{GS} pulls fewer source electrons into the N channel when L is shorter, which means R_{CH} rises. Although a shorter channel also increases conduction, the field effect of C_{GS} on R_{CH} is more intense. So R_{CH} is roughly 50% higher in saturation than deep in triode:

$$R_{CH}\Big|^{v_{GS}>v_{TN}}_{v_{DS}>v_{GST}} = \frac{V_{DS(SAT)}}{i_D} = \left(\frac{L_{CH}'}{W}\right)\left[\frac{v_{GS}-v_{TN}}{0.5K_N'(v_{GS}-v_{TN})^2}\right]$$

$$\approx \left(\frac{3}{2}\right)\left(\frac{L}{W}\right)\left[\frac{1}{0.5K_N'(v_{GS}-v_{TN})}\right] = 1.5 R_{CH}\Big|^{v_{GS}>v_{TN}}_{v_{DS}\ll v_{GST}}$$

(15)

Note R_{CH} is only the resistance of the channel. The total *drain–source resistance* R_{DS} or $\partial v_{DS}/\partial i_D$ in saturation is usually much greater than R_{CH} because i_D is largely insensitive to v_{DS}.

C. I–V Translation

Like in sub-threshold, i_D in inversion is sensitive to v_{DS} in triode and insensitive to v_{DS} in saturation like Fig. 21 shows. N⁺ regions can still reverse roles, so a positive v_{GD} and a negative v_{DS} establish a negative i_D that mirrors the i_D that positive v_{GS} and v_{DS} produce. i_D is zero when v_{GS} and v_{GD} are both negative.

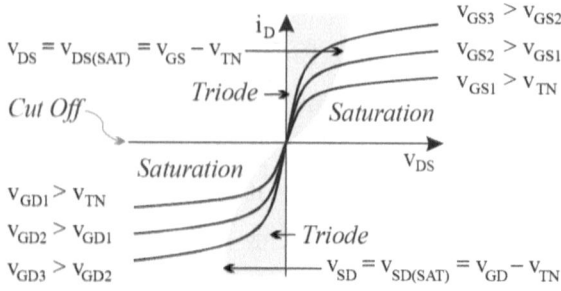

Fig. 21. Inverted N-channel MOSFET current.

The basic difference in inversion is that i_D is a squared translation of v_{GST}. The other difference is that i_D saturates when v_{DS} surpasses v_{GST}. So the $v_{DS(SAT)}$ boundary in Fig. 21 is a squared-root reflection of i_D:

$$v_{DS(SAT)} = v_{GS} - v_{TN} \approx \sqrt{\frac{2i_D}{(W/L)K_N'(1+\lambda_N v_{DS})}}\bigg|_{L \gg L_{MIN}} \approx \sqrt{\frac{2i_D}{(W/L)K_N'}}. \quad (16)$$

λ_N fades when L is much longer than the *minimum oxide length* possible L_{MIN} because L_{CH}' modulation is a small fraction of a long L.

Example 2: Determine $v_{DS(SAT)}$ when i_D is 100 µA, v_{DS} is 1 V, W is 10 µm, L is 180 nm, L_{OL} is 30 nm, K_N' is 200 µA/V², and λ_N is 2%.

Solution:

$$v_{DS(SAT)} \approx \sqrt{\frac{2i_D}{(W_{CH}/L_{CH})K_N'(1+\lambda_N v_{DS})}}$$

$$= \sqrt{\frac{2(100\mu)[180n - 2(30)]}{(10\mu)(200\mu)[1+(2\%)(1)]}} = 110 \text{ mV}$$

2.4. Body Effect

v_{GS} induces a field through C_{OX} that avails electrons for v_{DS} to pull. The body–source or bulk–source voltage v_{BS} also avails electrons the same way via C_{DEP}. So the P⁺ body terminal in Fig. 11 is a bottom gate.

A positive v_{BS} in NFETs avails electrons and a negative v_{BS} repels some of the electrons that v_{GS} avails. So v_{BS} effectively reduces v_{TN}:

$$v_{TN} = V_{TN0} + \gamma_N\left(\sqrt{2\psi_B - v_{BS}} - \sqrt{2\psi_B}\right). \quad (17)$$

v_{TN} is the baseline *zero-bias threshold* V_{TN0} when v_{BS} is zero.

MOSFETs

v_{TN} is the voltage that v_{GS} overcomes when inverting a channel. To invert in equal proportion, v_{GS} should not only negate the *surface–body barrier* ψ_B but also re-assert another ψ_B in the opposite direction. v_{BS} reduces this $2\psi_B$ translation. So v_{BS} alters v_{TN} by the amount that the *body-effect parameter* γ_N allows. This is the *body* or *bulk effect*, where ψ_S is $2\psi_B$ in inversion and γ_N and ψ_B can be 600 m√V and 300 mV.

Example 3: Determine v_{TN} when V_{TN0} is 400 mV, v_{BS} is −100 mV, γ_N is 600 m√V, and ψ_B is 300 mV.

Solution:

$$v_{TN} = V_{TN0} + \gamma_N \left(\sqrt{2\psi_B - v_{BS}} - \sqrt{2\psi_B} \right)$$

$$= 400m + (600m)\left(\sqrt{2(300m)+100m} - \sqrt{2(300m)} \right)$$

$$= 440 \text{ mV}$$

Note: A negative v_B pushes channel electrons back to v_S, so v_{GS} needs to overcome a higher v_{TN} to induce conduction.

Example 4: Determine v_{TN} when V_{TN0} is 400 mV, v_{BS} is 100 mV, γ_N is 600 m√V, and ψ_B is 300 mV.

Solution:

$$v_{TN} = V_{TN0} + \gamma_N \left(\sqrt{2\psi_B - v_{BS}} - \sqrt{2\psi_B} \right)$$

$$= 400m + (600m)\left(\sqrt{2(300m)-100m} - \sqrt{2(300m)} \right)$$

$$= 360 \text{ mV}$$

Note: A positive v_{BS} pulls v_S electrons into the channel region, so v_{GS} induces conduction more easily (with a lower v_{TN}).

2.5. Symbols

The NMOS is a four-terminal device with interchangeable v_S and v_D terminals that conduct i_D in Fig. 22 when v_{GS} and v_{DS} are positive. The two vertical lines at the gate symbolize the oxide capacitance that induces i_D. Static gate current i_G is zero because dc current into this C_{OX} is zero. So i_D is also the i_S that flows out of v_S. The arrow attaches to the v_S terminal that sets v_{GS} and points in the direction of i_S.

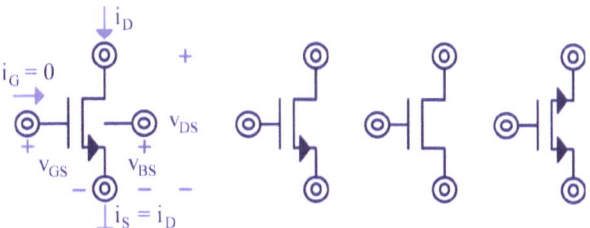

Fig. 22. N-channel MOSFET symbols.

The symbol sometimes excludes the body terminal to indicate other transistors share the same body. In these cases, independent access to the body is not possible. The arrow is also sometimes absent to show that source and drain terminals can reverse roles – this is typical in digital circuits. Although less of a convention, some switching power supplies add arrows to both terminals to confirm that i_D *will* at times reverse.

3. P-Channel MOSFETs

PFETs and NFETs function the same way, except PFETs rely on holes for conduction. So the PMOS structure is the complement of the NMOS. In PFETs, the $W_{CH} \times L_{OX}$ MOS capacitor in Fig. 23 hangs over N material and overlaps highly doped P$^+$ regions (across L_{OL}) that are L_{CH} apart. The P$^+$ terminals extend into the oxide region to connect them to the P channel that the oxide field forms. The highly doped N$^+$ region is an Ohmic surface contact to the body.

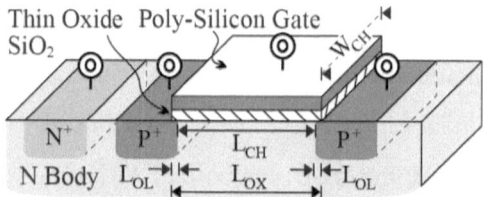

Fig. 23. P-channel MOSFET structure.

3.1. Accumulation: Cut Off

The N body usually connects to the most positive potential (*positive power supply* v_{DD} in Fig. 24) to keep P⁺N junctions from forward biasing. So the electrons and holes that diffuse recombine and deplete the regions near those junctions. Applying a positive v_G to the gate pulls electrons in the body toward the oxide. As a result, electrons accumulate near the semiconductor surface. This way, in accumulation, current does not flow, so the PFET is in cut off.

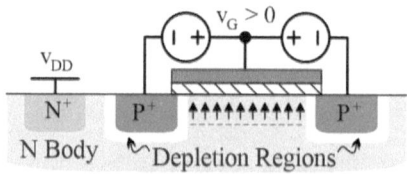

Fig. 24. P-channel MOSFET in accumulation and cut off.

3.2. Depletion: Sub-Threshold

Applying a negative v_G does the opposite: pushes electrons away from the semiconductor surface in Fig. 25, which depletes the region under the oxide of electrons. With fewer electrons with which to recombine, P⁺ holes diffuse farther before recombining. A negative v_G also presses bound electrons into their home sites, which eases hole movement. So v_G reduces the barrier voltage that keeps P⁺ holes from diffusing and extends their diffusion length.

The carrier density near the surface under the oxide falls to the point of becoming nearly intrinsic. E_F in this region in Fig. 26 is therefore nearly halfway across E_{BG}. But without any voltage between the P⁺ regions, current does not flow.

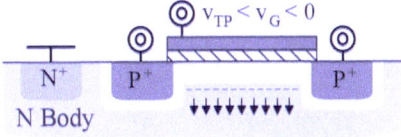

Fig. 25. P-channel MOSFET in depletion.

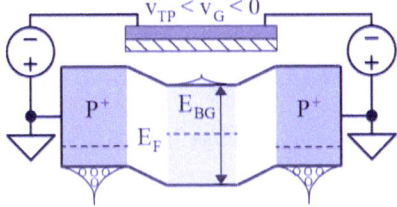

Fig. 26. Band diagram of P-channel MOSFET in depletion.

Decreasing the voltage of one of the P⁺ terminals elevates the barrier and expands the depletion region around that terminal (v_D in Fig. 27). The resulting field pulls diffusing holes into v_D. As E_{FLD} intensifies, more holes that would otherwise recombine reach v_D. Recombination nearly stops when v_D is 3 or 4 V_t's below v_S.

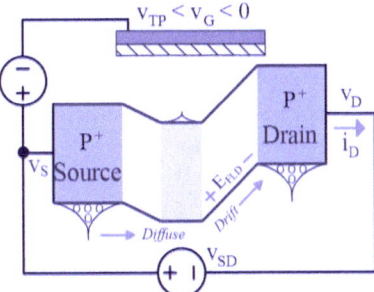

Fig. 27. Band diagram of P-channel MOSFET in sub-threshold.

The negative terminal is the drain because v_D outputs diffusing holes. The positive terminal is the source because v_S supplies these holes. The resulting flow of holes establishes an i_D that flows into v_S and out of v_D.

A. Triode

i_D increases with v_{SG} because a positive v_{SG} reduces the gate–source barrier. This rise is exponential because v_{SG} avails exponentially more holes than T_J avails with V_t:

$$i_D\Big|^{0<v_{SG}<|v_{TP}|} = \left(\frac{W}{L}\right)I_{SP}\exp\left[\left(\frac{C_{OX}"}{C_{OX}"+C_{DEP}"}\right)\left(\frac{V_{SGT}}{V_t}\right)\right]\left[1-\exp\left(\frac{-v_{SD}}{V_t}\right)\right]$$

$$= \left(\frac{W}{L}\right)C_{DEP}"\mu_P V_t^2 \exp\left(\frac{v_{SG}-|v_{TP}|}{n_I V_t}\right)\left[1-\frac{1}{\exp(v_{SD}/V_t)}\right] \quad .(18)$$

i_D also increases with v_{SD} because v_{SD} intensifies the field that pulls them to v_D. This i_D corresponds to triode because i_D is sensitive to v_{SG} and v_{SD}.

i_D is higher when W is wider, L is shorter, and the charge held in $C_{DEP}"$ is higher. These, *hole mobility* μ_P, and V_t set the baseline conductivity of i_D. I_{SP} incorporates the effects of $C_{DEP}"$, μ_P, and V_t.

The surface potential under the oxide is ultimately the voltage-divided fraction that v_G couples through C_{OX} across C_{DEP}. n_I models this voltage-divided reduction in gate drive v_{SGT} or $v_{SG} - |v_{TP}|$. Surface imperfections also impede diffusion, so n_I also accounts for these defects.

Holes do not drift as easily as electrons. Not surprisingly, native PFETs usually conduct less than native NFETs. Unfortunately, these PFETs oftentimes require 1.5 V or so for conduction, which can be too high for many applications. So process engineers implant a lightly doped layer of acceptor dopant atoms below the oxide. With some holes already present, conduction requires a lower v_{SG}. This implant therefore enhances the exponential effect of v_{SG}, which reduces the threshold voltage. Engineers define v_{TP} as a negative value because v_{GS} and v_{DS} in PFETs are correspondingly negative. But when using v_{SG} and v_{SD}, thinking of $|v_{TP}|$ is more insightful.

B. Saturation

i_D climbs with v_{SD} because v_{SD} intensifies the field, which reduces recombination. Diffusion length is so high in this quasi-intrinsic region that nearly 95% of the holes reach the other end when v_{SD} is $3V_t$. i_D becomes largely insensitive to v_{SD} past this point, saturating to

$$i_D \Big|_{v_{SD}>3V_t}^{0<v_{SG}<|v_{TP}|} \approx \left(\frac{W}{L}\right) I_{SP} \exp\left(\frac{v_{SG} - |v_{TP}|}{n_I V_t}\right). \tag{19}$$

C. I–V Translation

i_D in Fig. 28 is sensitive to v_{SD} in triode and insensitive to v_{SD} in saturation, when v_{SD} is greater than $3V_t$ or so. P$^+$ regions can change roles and conduct reverse current because the structure is symmetrical. So a positive v_{DG} and a negative v_{SD} establish a negative i_D that mirrors the i_D that positive v_{SG} and v_{SD} produce. i_D is zero in accumulation, when v_{SG} and v_{DG} are negative.

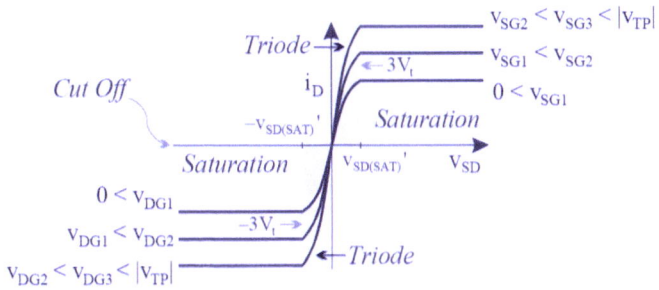

Fig. 28. Sub-threshold P-channel MOSFET current.

3.3. Inversion

A. Triode

Raising v_{SG} and v_{DG} first eliminates the barrier that keeps holes from flowing. Past this barrier point, the electric field across the oxide is high enough to pull holes from both P$^+$ regions in Fig. 29 toward the region below the gate. When v_{SG} and v_{DG} reach $|v_{TP}|$, these holes form a P channel with a carrier density that matches that of the original zero-bias N body. In other words, v_{SG} and v_{DG} "invert" a channel.

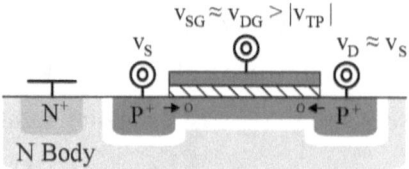

Fig. 29. Symmetrically inverted P-channel MOSFET in triode.

R_{CH} is high when L is long, W is narrow, and the conductivity that μ_P and $C_{OX}"$ in K_P' set is low. The channel dematerializes when v_{SG} and v_{DG} fall below $|v_{TP}|$. So R_{CH} spikes sharply when v_{SG} and v_{DG} reach $|v_{TP}|$:

$$R_{CH}\Big|_{v_{SD}<<v_{SGT}}^{|v_{SG}>|v_{TP}|} \approx \left(\frac{L}{W}\right)\left[\frac{1}{\mu_P C_{OX}"\left(v_{SG}-|v_{TP}|\right)}\right].$$

$$= \left(\frac{L}{W}\right)\left(\frac{1}{K_P' v_{SGT}}\right) \equiv \left(\frac{L}{W}\right) R_{SH}$$

(20)

R_{CH} is more accurate when v_{SG} and v_{DG} match and overcome $|v_{TP}|$. So v_{SG} and v_{DG} or $v_{SG} - v_{SD}$ are higher than $|v_{TP}|$ and v_{SD} is low. This means that gate drive $v_{SG} - |v_{TP}|$ or v_{SGT} is positive and v_{SD} is much lower than v_{SGT}. K_P' and v_{SGT} set the sheet resistivity R_{SH} of R_{CH}.

Example 5: Determine W so R_{CH} is no greater than 45 Ω when L is 180 nm, v_{SG} is 1.8 V, v_{SD} is 50 mV, μ_P is 14.4k mm²/V·s, t_{OX} is 12.5 nm, L_{OL} is 30 nm, and v_{TP} is –400 mV.

Solution:

$$C_{OX}" = \frac{\varepsilon_{OX}}{t_{OX}} = \frac{3.9\varepsilon_0}{t_{OX}} = \frac{3.9(8.845\times 10^{-12})}{12.5n} = 2.76 \text{ fF}/\mu m^2$$

$$K_P' = C_{OX}"\mu_P = (2.76m)(14.4m) = 40 \text{ }\mu A/V^2$$

$$v_{SD} = 50 \text{ mV} << v_{SG} - |v_{TP}| = 1.8 - 400m = 1.4 \text{ V}$$

$$R_{CH} \approx \left(\frac{L_{CH}}{W_{CH}}\right)\left[\frac{1}{K_P'(v_{SG} - |v_{TP}|)}\right]$$

$$\therefore = \left[\frac{180n - 2(30n)}{W_{CH}}\right]\left[\frac{1}{(40\mu)(1.8 - 400m)}\right] \leq 45\,\Omega$$

→ W ≡ $W_{CH} \geq 48$ μm

Note: W for the PFET is 5× higher than the NFET in Example 1 because μ_P is that much lower than μ_N.

---

v_{SD} across R_{CH} induces i_D. But since v_D opposes the inversion action of v_{DG} across half the channel, half of v_{SD} in Fig. 30 opposes the linear effects of v_{SG}. So R_{CH} rises and i_D falls with $0.5v_{SD}$:

$$i_D\Big|_{\substack{v_{SG} > |v_{TP}| \\ v_{SD} < v_{SGT}}} = \frac{v_{SD}}{R_{CH}} = v_{SD}\left(\frac{W}{L}\right)K_P'\left(v_{SG} - |v_{TP}| - \frac{v_{SD}}{2}\right). \qquad (21)$$

i_D climbs almost linearly with v_{SG} because an electric field (not diffusion) alters conductivity. This i_D corresponds to triode because i_D is sensitive to v_{SG} and v_{SD}.

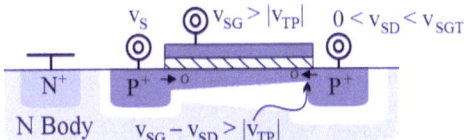

Fig. 30. Asymmetrically inverted P-channel MOSFET in triode.

B. Saturation

q_D near the edge of v_D's P⁺ region in Fig. 31 disappears when v_{DG} falls to $|v_{TP}|$. Since v_{DG} is $|v_{TP}|$ when this happens and v_{DG} is equivalent to $v_{SG} - v_{SD}$, the channel pinches when v_{SD} reaches $v_{SG} - |v_{TP}|$. So gate drive sets this saturation voltage $v_{SD(SAT)}$.

When v_{SD} overcomes this $v_{SD(SAT)}$, the channel extends to the fraction of L_{CH} that drops $v_{SD(SAT)}$: L_{CH}'. The depleted portion of L_{CH} drops what remains of v_{SD}: $v_{SD} - v_{SD(SAT)}$. i_D saturates because the voltage across the channel that L_{CH}' establishes is $v_{SD(SAT)}$, which is independent of v_{SD}:

$$i_D\Big|_{\substack{v_{SG}>|v_{TP}|\\v_{SD}>v_{SGT}}} = \frac{V_{SD(SAT)}}{R_{CH}} = v_{SD(SAT)}\left(\frac{W}{L_{CH}'}\right)K_P'\left(v_{SG}-|v_{TP}|-\frac{V_{SD(SAT)}}{2}\right)$$

$$\approx \left(\frac{W}{L}\right)\left(\frac{K_P'}{2}\right)\left(v_{SG}-|v_{TP}|\right)^2\left(1+\lambda_P v_{SD}\right) \qquad (22)$$

$$\equiv \left(\frac{W}{L}\right)\left(\frac{K_P'}{2}\right)V_{SGT}^2\left(1+\frac{V_{SD}}{V_{AP}}\right)$$

So i_D is sensitive to v_{SG} and largely insensitive to v_{SD}.

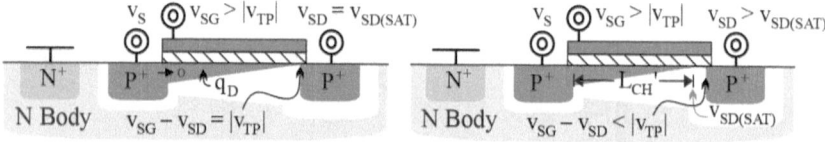

Fig. 31. Inverted P-channel MOSFET in saturation.

Since a lower v_D depletes more of the channel, L_{CH}' shrinks as v_{SD} rises. So R_{CH} falls and i_D rises with v_{SD}. λ_P models this L_{CH}' modulation with respect to the L that circuit designers define. The effects of v_{DS} fade as L lengthens because the variation becomes a smaller fraction of L, which means λ_P is lower and V_{AP} is higher.

Since C_{GS} falls with L_{CH}', v_{SG} pulls fewer source holes into the P channel when L is shorter, which means R_{CH} is higher. Although a shorter L_{CH}' also increases conduction, the field effect of C_{GS} is more intense. So R_{CH} is roughly 50% higher in saturation than in triode:

$$R_{CH}\Big|_{\substack{v_{SG}>|v_{TP}|\\v_{SD}>v_{SGT}}} = \frac{V_{SD(SAT)}}{i_D} = \left(\frac{L_{CH}'}{W}\right)\left[\frac{v_{SG}-|v_{TP}|}{0.5K_P'\left(v_{SG}-|v_{TP}|\right)^2}\right]$$

$$\approx \left(\frac{3}{2}\right)\left(\frac{L}{W}\right)\left[\frac{1}{0.5K_P'\left(v_{SG}-|v_{TP}|\right)}\right] = 1.5R_{CH}\Big|_{\substack{v_{SG}>|v_{TP}|\\v_{SD}>v_{SGT}}} \qquad (23)$$

R_{CH}, however, is only a fraction of R_{DS} in saturation. R_{DS} is usually much higher than R_{CH} because i_D is largely insensitive to v_{DS}.

C. I–V Translation

i_D in inversion is sensitive to v_{SD} in triode and insensitive to v_{SD} in saturation like Fig. 32 shows. Since P^+ regions can reverse roles, a positive v_{DG} and a negative v_{SD} establish a negative i_D that mirrors the i_D that positive v_{SG} and v_{SD} produce. i_D is zero when v_{SG} and v_{DG} are negative.

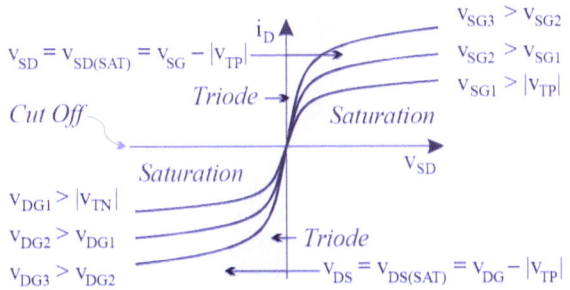

Fig. 32. Inverted P-channel MOSFET current.

In inversion, i_D is a squared translation of v_{SGT} and i_D saturates when v_{SD} overcomes v_{SGT}. So the $v_{SD(SAT)}$ boundary in Fig. 32 is a squared-root reflection of i_D:

$$v_{SD(SAT)} = v_{SG} - |v_{TP}| \approx \sqrt{\frac{2i_D}{(W/L)K_P'(1+\lambda_P v_{SD})}}\bigg|_{L \gg L_{MIN}} \approx \sqrt{\frac{2i_D}{(W/L)K_P'}} . \quad (24)$$

$\lambda_P v_{SD}$ fades when L is much longer than L_{MIN} because L_{CH}' modulation becomes a small fraction of L.

Example 6: Determine W so $v_{SD(SAT)}$ is no greater than 110 mV, i_D is 100 µA, v_{SD} is 1 V, L is 180 nm, L_{OL} = 30 nm, K_P' is 40 µA/V², and $1/\lambda_P$ is 50 V.

Solution:

$$V_{SD(SAT)} \approx \sqrt{\frac{2i_D}{(W_{CH}/L_{CH})K_P'(1+\lambda_P V_{SD})}}$$

$$= \sqrt{\frac{2(100\mu)[180n - 2(30n)]}{W_{CH}(40\mu)[1+(1/50)]}} \leq 110 \text{ mV}$$

$$\therefore \quad W \equiv W_{CH} \geq 49 \text{ μm}$$

Note: W is 5× wider than the NMOS in Example 2 because μ_P in K_P' is that much lower than μ_N in K_N'.

3.4. Body Effect

v_{SG} induces a field through C_{OX} that avails holes for v_{SD} to pull. The source–body or source–bulk voltage v_{SB} also avails holes the same way via C_{DEP}. So the N body terminal in Fig. 23 is a bottom gate.

A positive v_{SB} in the PMOS avails holes and a negative v_{SB} repels some of the holes that v_{SG} avails. So v_{SB} effectively reduces $|v_{TP}|$:

$$|V_{TP}| = |V_{TP0}| + \gamma_P \left(\sqrt{2\psi_B - v_{SB}} - \sqrt{2\psi_B} \right). \tag{25}$$

$|v_{TP}|$ is the baseline V_{TP0} when v_{SB} is zero.

$|v_{TP}|$ is the voltage that v_{SG} overcomes when inverting a channel. To invert in equal proportion, v_{SG} should not only negate the barrier ψ_B but also re-assert another ψ_B in the opposite direction. Except, v_{SB} reduces this $2\psi_B$ translation. And, v_{SB} alters $|v_{TP}|$ by the amount that γ_P allows.

3.5. Symbols

The PMOS is a four-terminal device with interchangeable v_S and v_D terminals that conduct i_D in Fig. 33 when v_{SG} and v_{SD} are positive. The two vertical lines at the gate symbolize the C_{OX} that induces i_D. i_G is zero because dc current into C_{OX} is zero. So i_D is also the i_S that flows into v_S. The arrow attaches to the v_S that sets v_{SG} and points in the direction of i_S.

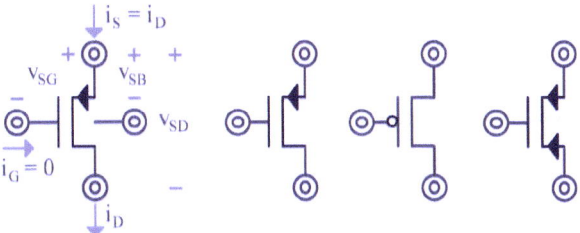

Fig. 33. P-channel MOSFET symbols.

The symbol sometimes excludes the body terminal to indicate other transistors share the same body. In these cases, independent access to the body is not possible. The arrow is also sometimes absent in digital circuits to show that source and drain terminals can reverse roles or on both terminals in switching power supplies to confirm that they will reverse roles. A "bubble" next to the gate distinguishes arrowless PFETs from NFETs. This indicates, like in a digital inverter, that PFETs "invert" the action of NFETs.

3.6. Unifying Convention

PFETs and NFETs function the same way. Accumulation, depletion, and inversion result when v_{GS} in NFETs and v_{SG} in PFETs are negative, positive and below v_{TN} and $|v_{TP}|$, and positive and above v_{TN} and $|v_{TP}|$. i_D saturates when v_{DS} in NFETs and v_{SD} in PFETs reach $v_{DS(SAT)}'$ and $v_{SD(SAT)}'$ in sub-threshold and $v_{DS(SAT)}$ and $v_{SD(SAT)}$ in inversion. v_{TN} and $|v_{TP}|$ fall when v_{BS} in NFETs and v_{SB} in PFETs are positive.

Everything that is positive in one is negative in the other. In PFETs, v_{GS}, v_{DS}, and v_{TP} are negative, and a negative v_{BS} diminishes v_{TP}. In other words, NFETs and PFETs deplete when v_{GS} reverses, invert when v_{GS} overcomes v_T's v_{TN} or v_{TP}, and saturate when v_{DS} overcomes the $v_{DS(SAT)}'$ or $v_{DS(SAT)}$ that $3V_t$ or $v_{GS} - v_T$ sets. So general v_{GS}, v_{DS}, v_{BS}, and v_T discussions apply to both: NFETs *and* PFETs.

4. Capacitances

The terminals of the MOSFET require time to charge and discharge. Most noticeably of these is the gate because t_{OX} is very thin (on the order of nanometers), so the resulting $C_{OX}"$ is substantial. Still, PN-junction capacitances at the source and drain also require charge and time to transition.

4.1. PN-Junction Capacitances

Body terminals normally connect to voltages that zero- or reverse-bias their PN junctions to sources and drains. A *reverse junction voltage* v_{JR} reinforces the diminishing effect of the *built-in potential* V_{BI} on *zero-bias junction capacitance* (per unit area) $C_{J0}"$. So *junction capacitance* C_J peaks to the C_{J0} that $C_{J0}"$ and *junction area* A_J set and v_{JR} reduces C_J to

$$C_J = \frac{C_{J0}"A_J}{\sqrt{\frac{V_{BI}+V_{JR}}{V_{BI}}}} = \frac{C_{J0}}{\sqrt{1+\frac{V_{JR}}{V_{BI}}}}, \quad (26)$$

where $C_{J0}"$ and V_{BI} depend strongly on the body's doping concentration.

Although source and drain geometries do not always match, A_J's are usually the same or comparable. v_{SB}, however, is usually lower than v_{DB}. So C_{SB} is normally higher than C_{DB}.

Although not necessarily so, doping densities in NFETs and PFETs usually differ. $C_{J0}"$ in NFETs is therefore different in PFETs. And, v_{JR}'s in NFETs are v_{SB} and v_{DB} and in PFETs are v_{BS} and v_{BD}.

4.2. Gate-Oxide Capacitances

Oxide capacitance decomposes into overlap and channel components. *Overlap capacitance* C_{OL} is the C_{OX} fraction that hangs over source and drain diffusions (along W_{CH} and across L_{OL} in Figs. 11 and 23):

$$C_{OL} = C_{OX}"W_{CH}L_{OL}. \quad (27)$$

Channel capacitance C_{CH} is the L_{CH} fraction of L_{OX} along W_{CH} that L_{OL}'s over source and drain diffusions exclude:

$$C_{CH} = C_{OX}"W_{CH}L_{CH} = C_{OX}"W_{CH}\left(L_{OX} - 2L_{OL}\right). \qquad (28)$$

But since the channel does not always extend across L_{CH}, C_{OX} decomposes into gate components differently across regions in Fig. 34.

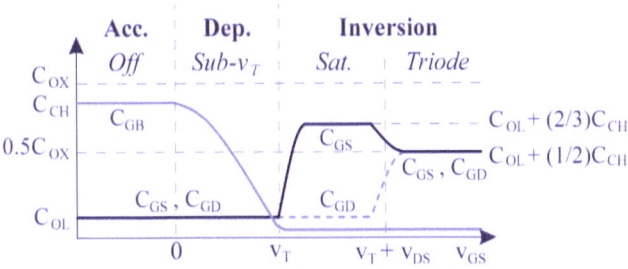

Fig. 34. Gate-oxide capacitances.

A. Cut Off

In accumulation, the region under the oxide in Figs. 12 and 24 is a good Ohmic contact to the body. This means that C_{CH} connects the gate to the body, so C_{GB} comprehends all of C_{CH} in Fig. 34. Since the gate overlaps the source and drain diffusions across L_{OL}, C_{GS} and C_{GD} only incorporate C_{OL}. So in cut off, the largest fraction of C_{OX} is in C_{GB}. C_{SB} and C_{DB} are the C_J's that their A_J's and v_{JR}'s set.

B. Sub-Threshold

When depleted, a depletion region separates the semiconductor surface (in Figs. 13 and 25) from the body. So C_{CH} stacks over C_{DEP} and C_{GB} is the series combination that results:

$$C_{GB} = C_{CH} \oplus C_{DEP} = \left(\frac{1}{C_{CH}} + \frac{1}{C_{DEP}}\right)^{-1}. \qquad (29)$$

Like parallel resistors, the series combination is lower than the smallest capacitance. C_{GB} is therefore lower than C_{CH} and C_{DEP}. And since the depletion region widens with higher v_{GS}'s, C_{GB} falls with C_{DEP} as v_{GS}

rises. With so little conduction in sub-threshold, C_{OL} is dominant in C_{GS} and C_{GD} and C_{JSB} and C_{JDB} are dominant in C_{SB} and C_{DB}.

C. Triode Inversion

When inverted in triode, a channel across the semiconductor surface connects v_S and v_D. C_{CH} therefore connects to v_G and the channel and C_{DEP} to the channel and v_B. Since v_S and v_D both connect to the channel, C_{GS} and C_{GD} share C_{CH} and C_{SB} and C_{DB} share C_{DEP}. So combined, C_{GS} and C_{GD} incorporate their C_{OL}'s plus matching C_{CH} halves:

$$C_{GS/GD(TRI)} = C_{OL} + 0.5C_{CH}, \qquad (30)$$

and C_{SB} and C_{DB} incorporate their C_J's plus matching C_{DEP} halves:

$$C_{SB/DB(TRI)} = C_{JSB/DB} + 0.5C_{DEP}. \qquad (31)$$

In other words, C_{GS} and C_{GD} carry equal C_{OX} fractions.

Note that FETs invert into triode when v_{GS} overcomes $v_T + v_{DS}$ because v_{DS} overcomes $v_{DS(SAT)}$'s $v_{GS} - v_T$ when this happens. Also notice that C_{GS}, C_{GD}, C_{SB}, and C_{DB} model all capacitances present. C_{GB} is a series combination of the C_{CH} that C_{GS} and C_{GD} model and C_{DEP}, the latter of which is very low in inversion.

D. Saturated Inversion

Inverted MOSFETs saturate when v_{GD} or $v_{GS} - v_{DS}$ drops below v_T, which happens when v_{GS} is between v_T and $v_T + v_{DS}$. In this mode, the channel (in Figs. 20 and 31) disconnects from v_D and shortens to L_{CH}'. So C_{DB} is C_{JDB} and C_{GD} is only the C_{OL} that L_{OL} sets. Since the channel still connects to v_S, C_{SB} carries C_{JSB} plus the C_{DEP} fraction that L_{CH}' sets and C_{GS} carries C_{OL} plus a similar fraction of C_{CH}:

$$C_{SB(SAT)} \approx C_{JSB} + (2/3)C_{DEP} \qquad (32)$$

$$C_{GS(SAT)} \approx C_{OL} + (2/3)C_{CH}, \qquad (33)$$

where the fraction is roughly two-thirds.

The C_{CH} and C_{DEP} fractions near v_D that L_{CH}' excludes are no longer in C_{GD} and C_{DB}. C_{GB} carries these fractions, but since C_{DEP} is so low, C_{GB} is usually negligible. In this mode, the largest fraction of C_{OX} is in C_{GS}.

E. Transition

C_{GS} and C_{GD} share C_{CH} equally when inverted in triode and v_{DS} is zero. As v_{DS} rises, the charge in the channel shifts towards the source. So C_{GS} acquires the corresponding C_{CH} fraction that C_{GD} in Fig. 35 loses. This continues until the channel pinches, when v_{DS} reaches $v_{DS(SAT)}$'s $v_{GS} - v_T$.

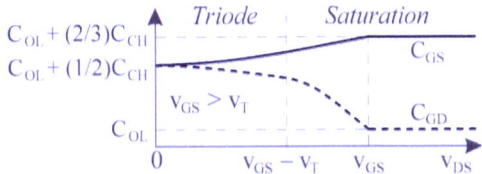

Fig. 35. Inverted gate–source and gate–drain capacitances.

As the channel recedes from the drain past $v_{DS(SAT)}$, the source and drain lose a small, but nevertheless growing fraction of C_{CH} to the body. As a result, C_{GS} acquires less of C_{CH} than C_{GD} loses as v_{DS} rises. This continues until C_{GD}'s fraction fades and C_{GS}'s share maxes to two–thirds. C_{GS} and C_{GD} reach their saturation limits this way when v_{DS} matches v_{GS}.

Example 7: Determine C_{GS} and C_{GD} in saturation when W is 10 µm, L is 180 nm, L_{OL} is 30 nm, and C_{OX}'' is 2.76 fF/µm².

Solution:

$$C_{OL} = C_{OX}''W_{CH}L_{OL} = (2.76m)(10\mu)(30n) = 0.83 \text{ fF}$$

$$C_{CH} = C_{OX}''W_{CH}L_{CH} = C_{OX}''W_{CH}(L_{OX} - 2L_{OL})$$
$$= (2.76m)(10\mu)[180n - 2(30n)] = 3.3 \text{ fF}$$

$$C_{GS} = C_{OL} + (2/3)C_{CH} = 0.83f + (2/3)(3.3f) = 3.0 \text{ fF}$$

$$C_{GD} = C_{OL} = 0.83 \text{ fF}$$

4.3. MOS Varactors

A. Bi-Modal

The MOS structure is fundamentally a parallel-plate capacitor with a thin dielectric. When used as a capacitor, paralleling all oxide components yields the highest capacitance. The PMOS in Fig. 36 combines all capacitive components by shorting v_S, v_B, and v_D terminals. This way, gate capacitance C_G incorporates C_{GS}, C_{GB}, and C_{GD}:

$$C_G = C_{GS} + C_{GB} + C_{GD} \leq 2C_{OL} + C_{CH} = C_{OX} = C_{OX}"W_{CH}L_{OX}. \quad (34)$$

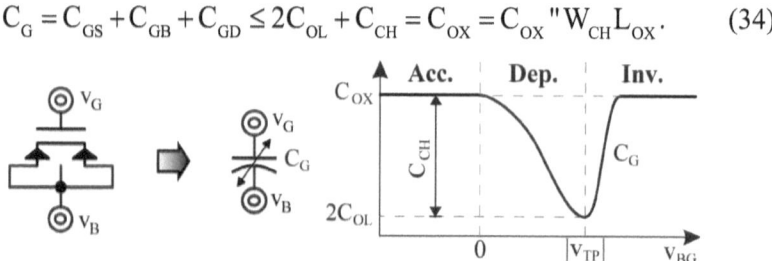

Fig. 36. Bi-modal P-channel MOSFET varactor.

When a positive v_{GB} (negative v_{BG}) accumulates electrons in the channel region, C_{GB} is C_{CH}, so C_G includes C_{GS} and C_{GD}'s $2C_{OL}$ and C_{GB}'s C_{CH}. When a v_{BG} that is greater than $|v_{TP}|$ inverts the channel that connects v_S and v_D, C_{GS} and C_{GD} each carry C_{OL} and half of C_{CH}. So even though C_{GB} is very low, C_{GS} and C_{GD} in C_G still carry $2C_{OL}$ and C_{CH}.

When a negative v_{GB} (positive v_{BG}) depletes the channel region without inverting it, C_{GB} becomes the series combination of C_{CH} and C_{DEP}. As v_{BG} climbs, v_G depletes more of the body, so C_{DEP} decreases, and with it, C_{GB}. So as v_{BG} reaches $|v_{TP}|$, C_G loses C_{GB}'s C_{CH}, falling from C_{OX} to the $2C_{OL}$ that C_{GS} and C_{GD} carry.

This structure is useful as a capacitor because C_G is high at C_{OX} when v_{BG} is negative and greater than $|v_{TP}|$. C_G is not a perfect variable capacitor or *varactor* because C_G is not monotonic with v_{BG}. A rise in v_{BG} does not always raise C_G because C_G is *bi-modal*. Incidentally, the PMOS symbol in Fig. 36 has two "source" arrows to indicate both P^+ diffusions supply holes when inverting the channel.

B. Inversion Mode

Disconnecting v_B from v_S and v_D removes C_{GB} from C_G. This way, the C_{CH} and C_{DEP} that C_{GB} in Fig. 34 adds in accumulation and depletion disappear from C_G in Fig. 37. So C_G's transition between $2C_{OL}$ and C_{OX} is now monotonic with v_{SG}. The drawback to this *inversion-mode* varactor is that the v_{SG} range that changes C_G is usually narrow. v_B connects to the highest potential to reverse body PN junctions to v_S and v_D.

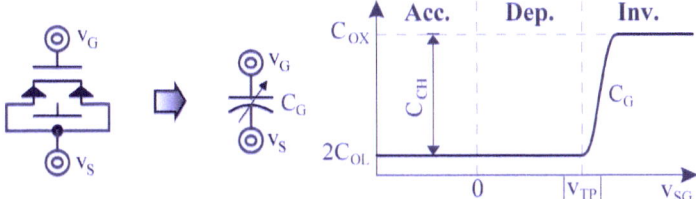

Fig. 37. Inversion-mode P-channel MOSFET varactor.

C. Accumulation Mode

C_{GB}'s transition in depletion in Fig. 34 is more gradual than C_{GS} and C_{GD}'s in inversion. A gradual transition is appealing because extending the voltage range that transitions capacitance is usually desirable in a varactor. So the purpose of the *accumulation-mode* structure in Fig. 38 is to eliminate the inversion mode from the bi-modal case.

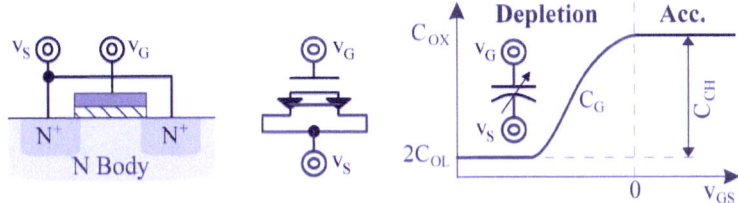

Fig. 38. Accumulation-mode N-channel MOSFET varactor.

The fundamental difference here is that the body is the same type of material as the source and drain. So the body connects the two N^+

terminals when v_{GS} is zero – the line across the source/drain terminals of the NMOS in Fig. 38 represents this connection. A positive v_{GS} reinforces the connection because it pulls and accumulates electrons under the oxide. And a negative v_G repels electrons and depletes the channel. This way, depletion reduces C_G from C_{OX} to $2C_{OL}$. Since no part of the structure can avail holes, the channel never inverts.

D. Variations

The varactors in Figs. 36–38 are P-, P-, and N-channel FETs because most fabrication technologies stack NFETs on P substrates. This means, independent P-type body terminals are not accessible. If they were, N-, N-, and P-channel devices would also be possible. These varactors are useful in *voltage-controlled oscillators* (VCO) because their voltages can adjust the frequency that their capacitances establish.

4.4. MOS Diodes

Another useful application of MOSFETs in power supplies and other analog circuits are as diodes. Key to their realization is C_{GS}, first as a stand-alone component and then as the agent that induces conduction. The other critical element is an inverting feedback loop.

A. Diode Connection

In Fig. 39, the drain–gate connections close a feedback loop around C_{GS} and the channel that induce the MOSFETs to operate like diodes. The NMOS, for example, is off when v_S is grounded and v_G is low. But when a circuit feeds *input current* i_{IN} into the drain–gate node, i_{IN} charges C_{GS}. When v_{GS} overcomes v_T, channel current i_D begins to sink some of i_{IN}. The rest of i_{IN} continues to charge C_{GS} until i_D is able to sink all of i_{IN}, at which point C_{GS} stops charging. So i_{IN} "forward-biases" the NFET.

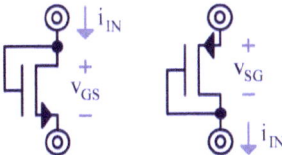

Fig. 39. N- and P-channel MOSFET diodes.

Since v_{DS} equals v_{GS}, v_{DS} is usually greater than the $3V_t$ that sets $v_{DS(SAT)}'$. So if i_{IN} is not high enough to invert a channel, the MOSFET saturates in sub-threshold. This means that v_{GS} is a logarithmic translation of i_{IN} that v_T offsets:

$$v_{GS} \approx v_T + n_I V_t \ln\left[\frac{i_{IN}}{(W/L)I_{SN}}\right] < v_T. \quad (35)$$

When i_{IN} inverts a channel, v_{DS} is also greater than the $v_{GS} - v_T$ that sets $v_{DS(SAT)}$ because v_{DS} matches v_{GS}. So the MOSFET inverts into saturation. v_{GS} is therefore not only a $v_{DS(SAT)}$ reflection but also a square-root translation of i_{IN} that v_T offsets:

$$v_{GS} = v_T + v_{DS(SAT)}$$

$$\approx V_{TN0} + \gamma_N\left(\sqrt{2\psi_B - v_{BS}} - \sqrt{2\psi_B}\right) + \sqrt{\frac{2i_{IN}}{(W/L)K'(1+\lambda v_{DS})}}. \quad (36)$$

Note that body effect alters v_T, L_{CH} modulation alters $v_{DS(SAT)}$, and L_{CH} modulation is higher when channels are shorter.

When i_{IN} is no longer present, i_D discharges C_{GS} to v_T in inversion and below v_T in sub-threshold. So i_{IN} activates and cuts off *diode-connected* MOSFETs like i_{IN} would a junction diode. The only difference is that a diode drops a logarithmic translation of i_{IN} and MOSFETs drop a squared-root translation offset by v_T. *MOS diodes* that drop 300–500 mV, however, are often preferable because i_{IN} burns less power across 300–500 mV than across the 600–800 mV that diodes establish.

Pulling i_{IN} from v_S when the gate–drain terminals connect to a voltage that is above ground similarly charges C_{GS} until i_D conducts i_{IN}. The PMOS does the same when a circuit feeds or pulls i_{IN} to v_S or from the gate–drain node. The only difference is their v_{GS} because v_{TN} and μ_N do not match $|v_{TP}|$ and μ_P.

B. Diode Action

Sometimes this *diode action* results from an *implicit* connection. One example is when other circuit components connect gate and drain terminals together. Another example is pulling i_{IN} from v_S in Fig. 40 when a voltage or a large capacitor holds v_G because i_{IN} charges C_{GS} until i_S (and whatever connects to v_D) supplies i_{IN}. The PMOS does the same when a circuit feeds i_{IN} to v_S: i_{IN} charges C_{GS} until i_S (and whatever connects to v_D) sinks i_{IN}.

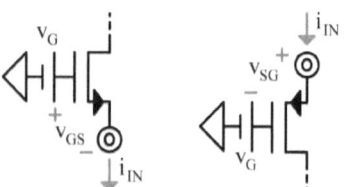

Fig. 40. Implicit N- and P-channel MOS diode action.

Example 8: Determine v_S for an NMOSFET when i_{IN} pulls 100 µA from v_S, W is 10 µm, L is 180 nm, L_{OL} is 30 nm, K_N' is 200 µA/V², V_{TN0} is 400 mV, γ_N is 600 m√V, ψ_B is 300 mV, λ_N is 10%, and v_B, v_G, and v_D are zero.

Solution:

$L_{CH} = L - 2L_{OL} = 180n - 2(30n) = 120$ nm

$v_{DS} = v_D - v_S = v_{GS} = v_G - v_S = v_{BS} = v_B - v_S = 0 - v_S = -v_S$

$$v_{TN} = V_{TN0} + \gamma_N \left(\sqrt{2\psi_B - v_{BS}} - \sqrt{2\psi_B} \right)$$

$$= 400\text{m} + 600\text{m} \left(\sqrt{2(300\text{m}) + v_S} - \sqrt{2(300\text{m})} \right)$$

$$V_{DS(SAT)} = \sqrt{\frac{2(100\mu)}{(10\mu/120\text{n})(200\mu)\left[1 - 10\% v_S\right]}}$$

$v_{GS} = v_{TN} + v_{DS(SAT)} = 340 \text{ mV} \rightarrow v_S = -340 \text{ mV}$

Note: Example 11 shows why the FET is inverted. v_{BS} reduces v_T more than $V_{DS(SAT)}$ raises v_{GS}. λ_N suppresses v_{DS} more than γ_N suppresses v_{BS}, so neglecting L_{CH} modulation yields a similar v_{GS}. These implicit diode conditions are typical for ground NMOSFETs in switched-inductor power supplies.

5. Short Channels

Smaller geometries are appealing in three basic ways. First, they occupy less *silicon area*, so FETs cost less and microchips can fit more circuits. Second, capacitances are lower, so FETs require less charge power and less time to transition. And third, electric fields are stronger, so conduction requires less voltage, and as a result, less power.

Unfortunately, geometric reductions and stronger electric fields produce unappealing effects in conductivity and noise that are not always easy to model or counter. These usually surface when L_{CH} is comparable to the *depletion widths* d_W around the source and drain terminals. This is why sub-micron devices suffer from *short-channel effects* that fade and disappear in longer-channel devices.

5.1. Drain-Induced Barrier Lowering

In NFETs, increasing v_D in Fig. 41 pulls N^+ electrons away from the PN junction towards v_D's contact and pushes P-body holes away into the body. So v_D's depletion region expands in all directions. When L_{CH} is comparable to d_W's, v_D's depletion region can extend and merge into v_S's. The merging of two depletion regions this way is *punch through*.

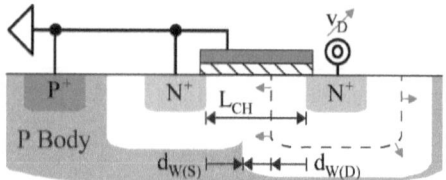

Fig. 41. Drain-induced punch-through.

v_D's field is so close to v_S that it pushes holes away from the channel region near v_S and loosens nearby electrons. So raising v_{DS} helps deplete and invert the channel. This way, a lower v_{GS} can more easily induce conduction. This means, v_{DS} reduces the barrier voltage v_B in Fig. 42 that keeps electrons from diffusing. This *drain-induced barrier lowering* (DIBL) effectively reduces v_T.

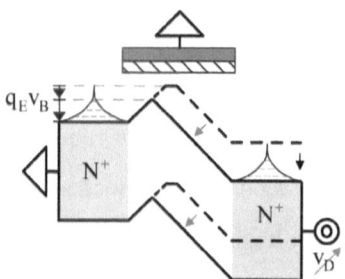

Fig. 42. Drain-induced barrier lowering in an N-channel MOSFET.

This is a problem because v_{DS} can induce current flow with zero v_{GS}. So v_G may not be able to cut the NFET off, which is another way of saying v_G can lose control of the NFET. Unfortunately, this effect is not static because v_D changes with time.

In PFETs, v_D's field is so close to v_S that it pushes electrons away from the oxide region near v_S and presses nearby valence electrons into their home sites. Holes can therefore drift more easily. So PFETs also suffer a dynamic reduction in v_T when v_D falls.

A. Thinner Oxide

The surface potential is ultimately the result of capacitor coupling from v_G, v_B (via the body effect), and v_D (with DIBL). So in the absence of v_G and v_B, ψ_S in Fig. 43 is the voltage-divided fraction that v_D's depletion capacitance C_{JD} to the channel couples across v_G's C_{OX}, v_S's C_{JS}, and v_B's C_{JB}. DIBL is noticeable because short L_{CH}'s increase C_{JD}'s coupling.

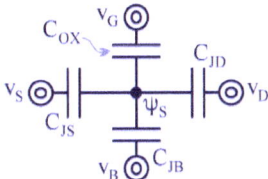

Fig. 43. Channel coupling components.

The effect of C_{OX}, C_{JS}, and C_{JB} is to shunt C_{JD}'s coupling. So raising C_{OX} reduces DIBL. This is one of the driving reasons why engineers scale t_{OX} with L_{OX}. Another reason is higher current density i_D/W_{CH} and v_{GS}-to-i_D gain because C_{OX}'' in K' climbs with reductions in t_{OX}. Reducing t_{OX} from 25 to 5 nm, for example, can suppress the 250-mV reduction in v_T that 100 mV across v_{DS} can produce when L_{OX} is 40 nm.

5.2. Gate–Channel Field

A. Surface Scattering

Thinner t_{OX}'s intensify vertical gate–channel fields. So on their way to the drain, carriers accelerate and collide with the oxide on the surface of the semiconductor in Fig. 44 more often and with greater force. This scattering effect reduces *surface mobility* and produces noise in i_D. *Surface scattering* intensifies as L_{CH} and t_{OX} scale down.

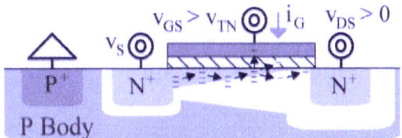

Fig. 44. Surface scattering and hot-electron injection in the NMOS.

B. Hot-Electron Injection

Positive gate–channel fields energize, accelerate, and direct N-channel electrons into the oxide. When charged with sufficient *kinetic energy* E_K, these *hot electrons* can break into or tunnel through the oxide. Electrons that break into and stay in the lattice (in Fig. 44) leave the oxide negatively charged. So over time, a higher v_{GS} is necessary to deplete and invert the channel, which means v_{TN} rises. And electrons that tunnel through the thin oxide establish a gate current i_G. PFETs are largely immune to *hot-electron injection* because their v_{GS}'s are usually negative.

C. Oxide-Surface Ejections

In repelling electrons, negative gate–channel fields weaken electron bonds along the silicon–oxide interface. When sustained and at elevated temperatures, they break silicon–hydrogen (Si–H) bonds and dispel negatively charged H atoms into the body. These atoms leave behind positively charged "hole" traps that counter the action of negative v_{GS}'s.

So PFETs need a higher v_{SG} to deplete and invert the channel, which means $|v_{TP}|$ is higher. But as v_{SG} weakens the field, electrons repopulate holes, so v_{TP} recovers. v_{TP} therefore fluctuates as v_{SG} stresses and relaxes the oxide. This *negative bias temperature instability* (NBTI) is less prevalent in NFETs because they are less prone to negative v_{GS}'s.

D. Fringing Fields

Shrinking planar dimensions enhance the effects of fringing fields along the periphery of the gate-oxide region in Fig. 45 on the channel. These field lines deplete and invert space outside the W_{CH} that source and drain diffusions define, extending the width of the channel. This W_{CH} variation ΔW_{CH} is negligible in larger FETs, but substantial in sub-micron devices.

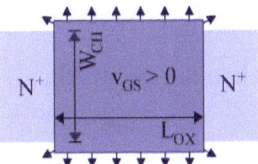

Fig. 45. Fringing electrics-field lines around an N-channel MOSFET.

5.3. Source–Drain Field

A. Velocity Saturation

Electron velocity v_E in the *conduction band* scales linearly with voltage up to 100 km/s or so. *Hole velocity* v_H scales similarly, but saturates at 60 km/s or so. $v_{H(SAT)}$ is lower than $v_{E(SAT)}$ because the *valence electrons* that shift holes are more tightly bound to their home sites than electrons in the conduction band. Their mobility ultimately determines the *critical electric fields* E_C that accelerate them to these levels:

$$E_{CN} = \frac{v_{E(SAT)}}{\mu_N} \quad (37)$$

and

$$E_{CP} = \frac{v_{H(SAT)}}{\mu_P}. \quad (38)$$

E_{CN} and E_{CP} in silicon are 1.4 and 4.2 V/μm at room temperature. E_{CP} is higher because μ_P is lower than μ_N more than $v_{H(SAT)}$ is lower than $v_{E(SAT)}$.

In triode inversion, i_D scales with v_{DS} until v_{DS} saturates v_E or v_H. The *velocity saturation voltage* $v_{DS(SAT)}"$ that E_C across L_{CH} sets is

$$v_{DS(SAT)}" = E_C L_{CH}. \quad (39)$$

v_E and v_H therefore saturate when v_{DS} across a 1-μm channel reaches 1.4 and 4.2 V. But if the $v_{DS(SAT)}$ that $v_{GS} - v_T$ sets is less than 1.4 V in 1-μm NFETs and 4.2 V in 1-μm PFETs, v_E and v_H do not saturate.

$v_{DS(SAT)}"$ for sub-micron channels can be so low that i_D can saturate before v_{GD} pinches the channel at $v_{DS(SAT)}$. i_D in Fig. 46, for example, scales with v_{DS} in triode inversion until v_{DS} reaches $v_{GS} - v_T$ when L_{CH} is long and $E_C L_{CH}$ when L_{CH} is short. Normally, MOSFETs begin to suffer from velocity saturation when L_{CH} is less than 1 μm.

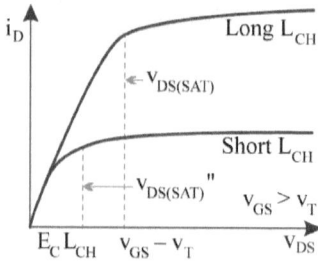

Fig. 46. Velocity saturation and pinch-off effects in inversion.

Example 9: Determine $v_{DS(SAT)}''$ and $v_{SD(SAT)}''$ when L_{CH} is 180 nm.
Solution:

$$v_{DS(SAT)}'' = E_{CN} L_{CH} = (1.4/\mu)(180n) = 250 \text{ mV}$$

$$v_{SD(SAT)}'' = E_{CP} L_{CH} = (4.2/\mu)(180n) = 760 \text{ mV}$$

Note: $v_{DS(SAT)}''$ and $v_{SD(SAT)}''$ are over the 130-mV $v_{DS(SAT)}$ and $v_{SD(SAT)}$ that 10- and 50-μm wide and 180-nm long N and P MOSFETs set with 100 μA (from previous examples), so v_{GD} pinches their channels before carrier velocities saturate.

B. Impact Ionization

Electrons gain speed and E_K when source–drain fields intensify. E_K can be so great that these *hot electrons* can collide and liberate otherwise immobile electrons from their home sites. Engineers call this process *impact ionization* because atoms ionize on impact.

Each energized electron can free an electron that avails a hole. Two electrons can then gain enough E_K to liberate another two *electron–hole pairs* (EHP) like Fig. 47 shows. This v_{DS}-induced process repeats, multiplies, and grows *avalanche* fashion.

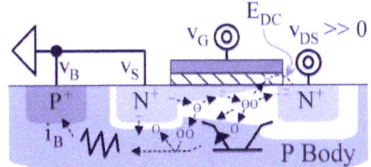

Fig. 47. Impact ionization, avalanche, and hot-electron injection.

Impact ionization is a problem in MOSFETs because, while electrons in NFETs scatter toward the positively charged drain, holes flow into the negatively charged body. Since the body is moderately doped, body resistance drops a voltage that forward-biases v_{BS}'s PN junction. So v_S's N$^+$ electrons also diffuse across the junction into v_B.

Forward-biasing this bulk–source junction activates the lateral NPN across the source–body–drain regions. When L_{CH} is very short, the N$^+$ drain "collects" most of the electrons that the N$^+$ source "emits" into the channel. So the NPN draws *body current* i_B from v_B and conducts i_D. In other words, the NPN can short the NFET with zero v_{GS}.

PFETs suffer the same effect. Except, holes scatter toward the negatively charged drain and electrons flow into the positively charged body. The resulting i_B drops a voltage that forward-biases v_{SB}'s PN junction and activates the lateral PNP across the source–body–drain regions. So even with zero v_{SG}, the PNP can short the PFET.

C. Arcing Field

Shrinking planar dimensions enhance the effects of arcing channel–drain field E_{DC} lines that pass through the gate oxide near the drain in Fig. 47. This E_{DC} intensifies as L_{OX} shortens, v_D rises, and the $v_{DS(SAT)}$ across the channel that $v_{GS} - v_T$ sets falls. L_{OX} in sub-micron devices is so short and lateral and arcing fields so intense as a result that N-channel electrons can gain enough E_K to break into the oxide and stay there. These hot electrons charge the oxide, so v_{TN} increases.

D. Lightly-Doped Drain

In triode inversion, the channel drops v_{DS} across L_{CH}. When pinched, the channel drops the $v_{DS(SAT)}$ that $v_{GS} - v_T$ sets across L_{CH}'. So the depletion region between the drain and channel drops the remainder $v_{DS} - v_{DS(SAT)}$ or $v_{DG} + v_T$ across $L_{CH} - L_{CH}'$. Of these, $L_{CH} - L_{CH}'$ is the shortest distance. So the most intense field usually results across this short drift space. This field intensifies when L_{CH} shortens and v_{DG} climbs.

Outside of lengthening L_{CH} and reducing v_{DG}, the only other way of weakening this field is by extending v_D's depletion length to the channel. The *lightly doped drain* (LDD) regions in Fig. 48 do this by depleting farther into the drain. Because with fewer carriers, v_D depletes farther into the LDD region. This way, the resulting drift length is longer than $L_{CH} - L_{CH}'$.

Fig. 48. Lightly-doped-drain MOSFETs.

For this, engineers first implant LDD dopants into the silicon without the oxide spacers shown. Then, with the spacers, implanting more dopants forms the highly doped regions without altering the LDDs. Both source and drain regions receive LDDs because they swap roles when i_D reverses. Plus, only the voltage at the terminal that acts as the drain can deplete, leaving the source largely intact. This practice reduces impact ionization, avalanche, and electron injection. And since LDDs are shallow, v_D depletes less channel space, so DIBL is also lower.

6. Other Considerations

6.1. Weak Inversion

Not surprisingly, inverting the channel does not keep carriers from diffusing across source–drain terminals. In fact, v_{DS} induces carriers to both diffuse *and* drift. And v_{GS} determines to what extent.

Deep in sub-threshold, when v_{GS} is well below v_T in Fig. 49, *drift current* i_{FLD} is so low that *diffusion current* i_{DIF} dominates i_D. And i_{FLD} is so high in *strong inversion*, when v_{GS} is well above v_T, that i_{FLD} dwarfs i_{DIF}. Near v_T, i_{DIF} and i_{FLD} are comparable. In this context, v_T is the v_{GS} that produces matching i_{DIF} and i_{FLD} components. So in *weak inversion*, as the channel forms, i_D reflects both conduction mechanisms.

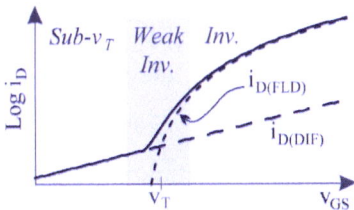

Fig. 49. Drain current as MOSFET channel forms.

Solving accurate sub-threshold and inversion expressions for every single transistor across time in a system that incorporates thousands, if not billions of transistors can be time-consuming for a computer. To save computing time, computer platforms model one region well and estimate the other. Or if they model both regions well, they approximate weak inversion. So for more predictable and reliable operation, engineers often design MOSFETs to operate deep in sub-threshold or in strong inversion.

A. Voltage Bias

Digital circuits and switching power supplies normally use MOSFETs as switches. These FETs close and open into the on and off states that v_{GS} establishes. The v_{GS} that closes FETs is usually much greater than v_{DS} because v_{DS} is only millivolts after FETs close. FETs open when v_{GS} is zero. So these FETs switch between triode and cut off, and saturate only during transitions. When closed, they are in sub-threshold when v_{GS} cannot overcome v_T and in inversion when v_{GS} can.

To assert the least and most resistive on and off states, circuit designers normally apply the highest and lowest v_{GS}'s possible. When voltages higher than v_T are not available, MOSFETs cannot invert a channel when they close. So they switch between sub-threshold and cut off. Inversion is only possible when voltages higher than v_T are available, in which case *on resistance* R_{ON} is R_{CH} in triode inversion.

B. Current Bias

Amplifiers and linear power supplies normally bias FETs at particular v_{GS}–v_{BS}–v_{DS}–i_D settings. Then, they vary one or two of these variables and use variations in one or two of the others to drive other circuits into action. v_{GS} is normally 0.5–2 V and v_{DS} is higher than 300 mV. So for the most part, these FETs operate in saturation or on the edge of saturation. In other words, triode operation is less likely in these applications.

Analog performance is sensitive to bias conditions, so current and voltage settings should be predictable and stable. Except predicting and stabilizing a v_{GS}-defined i_D is difficult because i_D is sensitive to v_{GS} and v_T, v_{GS} is noisy, and v_T varies widely across fabrication corners. Defining v_{GS} with i_D is usually better because the exponential and quadratic v_{GS} terms that set i_D in sub-threshold and inversion suppress the v_{GS} variations that i_D produces. This is why i_D (instead of v_{GS}) is usually one of the design parameters used to bias transistors in analog circuits.

Inverting a MOSFET, however, is not an option when voltages higher than v_T are not available. But it is otherwise. Still, *analog designers* often prefer sub-threshold for low power consumption because both voltages and currents are low. They like inversion for high speed because the higher currents that inversion induces charge and discharge capacitances quicker.

Since v_{GS} is an indirect logarithmic or squared-root translation of i_D that v_T shifts, ensuring i_D-biased MOSFETs are in sub-threshold or inversion is not straightforward. Luckily, what usually matters most to analog design engineers is predictable *small-signal performance*. And this hinges on *small-signal transconductance* g_m because g_m translates small-signal variations in v_{GS} on i_D and *vice versa*.

Small v_{GS} signals are so much smaller than v_{GS} that a linear slope translation can approximate their effect on i_D fairly well. This g_m slope is i_D's first partial derivative $\partial i_D/\partial v_{GS}$ with respect to v_{GS}. Since v_{GS} is within an exponential in i_D in sub-threshold, $\partial i_D/\partial v_{GS}$ matches i_D, but with v_{GS}'s $1/n_I V_t$ coefficient as a multiplier:

$$g_m\Big|_{V_{DS}>3V_t}^{0<v_{GS}<v_T} \equiv \frac{\partial i_D}{\partial v_{GS}} \approx \left(\frac{1}{n_I V_t}\right)\left(\frac{W}{L}\right) I_S \exp\left(\frac{v_{GS}-v_T}{n_I V_t}\right) \approx \frac{i_D}{n_I V_t} \approx \frac{I_D}{n_I V_t}, \quad (40)$$

where variations in i_D are so much smaller than i_D's static (non-varying) component I_D that i_D reduces to I_D. In inversion, $\partial i_D/\partial v_{GS}$ loses i_D's quadratic effect, so g_m is a square-root translation of I_D:

$$g_m\Big|_{V_{DS}>v_{GST}}^{v_{GS}>v_T} \equiv \frac{\partial i_D}{\partial v_{GS}}$$

$$= \left(\frac{W}{L}\right) K'(v_{GS}-v_T)(1+\lambda v_{DS}) \quad , \quad (41)$$

$$= \sqrt{2i_D K'(W/L)(1+\lambda v_{DS})} \approx \sqrt{2I_D K'(W/L)}$$

where λv_{DS}'s channel-length modulation effect fades when L is long.

Note that g_m rises with W/L in inversion, but not in sub-threshold. And a squared-root suppresses i_D in inversion, but not in sub-threshold. So with the same i_D, g_m in Fig. 50 climbs with W/L until g_m maxes in sub-threshold. From this perspective, v_T is the v_{GS} that maxes g_m in inversion to the $i_D/n_I V_t$ that sets g_m in sub-threshold:

$$g_{m(MAX)}\Big|_{\substack{v_{GS}>v_T \\ v_{DS}>v_{GST}}} = \sqrt{2i_D K'(W/L)_X (1+\lambda v_{DS})} = g_m\Big|_{\substack{0<v_{GS}<v_T \\ v_{DS}>3V_t}} \approx \frac{i_D}{n_I V_t}. \quad (42)$$

This happens when W/L is $(W/L)_X$ and $v_{DS(SAT)}$ is $2n_I V_t$:

$$(W/L)_X \approx \frac{i_D}{2n_I^2 V_t^2 K'(1+\lambda v_{DS})} \quad (43)$$

and $\quad v_{DS(SAT)}\Big|_{\substack{v_{GS}>v_T \\ (W/L)_X}} = \sqrt{\dfrac{2i_D}{(W/L)_X K'(1+\lambda v_{DS})}} = 2n_I V_t. \quad (44)$

In other words, MOSFETs invert when the $v_{DS(SAT)}$ that W/L sets at a particular i_D is greater than $2n_I V_t$.

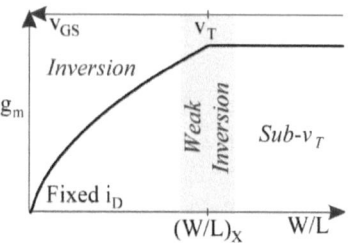

Fig. 50. Small-signal transconductance across regions.

This is handy because i_D and W/L are the variables that engineers use to design circuits. But since MOSFETs invert weakly near $2n_I V_t$ and weak-inversion models are not always accurate, adding a ±25% guard-band is prudent. So $v_{DS(SAT)}$ should be less than $1.5n_I V_t$ for deep sub-threshold and greater than $2.5n_I V_t$ for strong inversion.

Example 10: Determine the W/L that ensures a PMOSFET is in sub-threshold when i_D is 1 µA, T_J is 27 °C, K_P' is 40 µA/V² at 27 °C, n_I is 1.75, *Boltzmann's constant* K_B is 1.38×10^{-23} J/K, *electronic charge* q_E is 1.60×10^{-19}, and L is $10L_{MIN}$.

Solution:

$$V_t = \frac{K_B T_J}{q_E} = \frac{(1.38)(10^{-23})(300)}{(1.60)(10^{-19})} = 26 \text{ mV}$$

$L \gg L_{MIN}$

$$\therefore v_{DS(SAT)} \approx \sqrt{\frac{2i_D}{(W/L)K_P'}} = \sqrt{\frac{2(1\mu)}{(W/L)(40\mu)}}$$

$$\leq 1.5 n_I V_t = 1.5(1.75)(26m) = 68 \text{ mV}$$

$\rightarrow$ W/L $\geq$ 11 $\therefore$ W/L $\equiv$ 15

Note: K_P' typically falls with T_J, so although not to the same extent, $v_{DS(SAT)}$ tends to track V_t's rise with T_J. So this W/L may keep the PFET in sub-threshold across temperature.

Example 11: Determine if the NMOSFET in Example 8 is inverted when n_I is 1.75.

Solution:

$$v_{DS(SAT)} = \sqrt{\frac{2i_{IN}}{(W_{CH}/L_{CH})K_N'(1+\lambda_N v_{DS})}}$$

$$\approx \sqrt{\frac{2i_{IN}}{(W_{CH}/L_{CH})K_N'}} = \sqrt{\frac{2(100\mu)}{(10\mu/120n)(200\mu)}}$$

$$= 110 \text{ mV} > 2n_I V_t = 2(1.75)(25.6m) = 90 \text{ mV}$$

$\therefore$ Inverted

6.2. Junction Isolation

Basic *complementary MOS* (CMOS) technologies integrate N- and P-channel FETs into one silicon substrate. More sophisticated technologies integrate and optimize other components. But they cost more because they require additional fabrication steps. Still, the benefits of an expanded list of optimized components often outweigh the added cost.

Sharing a common silicon substrate requires electronic isolation from the substrate. This means that no circuit component should inject or draw *substrate current* i_{SUB}. Luckily, reverse-biasing all PN junctions to the substrate does this. So when ground is the most negative potential, the P substrate in Fig. 51 should connect to ground. N substrates should similarly connect to the most positive potential, which is often v_{DD}.

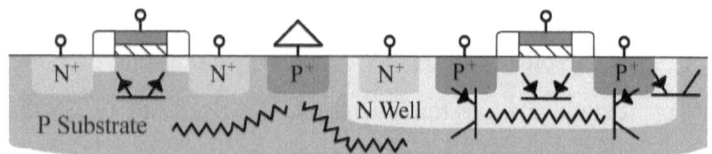

Fig. 51. Junction-isolated single-well P-substrate CMOS FETs.

Junction isolation is not perfect, however. For one, every component in the microchip incorporates junction capacitance C_J to the substrate, which requires current and time to charge and discharge. C_J's also couple noise into the substrate, and through the substrate, into other devices. Plus, unintended breakdown and transient excursions can forward-bias substrate PN junctions and inject i_{SUB} into the shared substrate. Still, junction isolation is straightforward and cost-effective.

A. Channel BJTs

Oppositely doped source–body–drain diffusions also realize source–drain BJTs. Source, body, and drain terminals are the emitter, base, and collector because sources supply the electrons in NFETs and holes in PFETs that drains collect. *Base–collector current gain* β_0 can be high because the *effective base width* w_B across sub-micron channels is short.

Channel BJTs in the substrate are usually in cut off because substrate PN junctions to source and drain diffusions are usually reverse-biased. Source–drain BJTs in wells also cut off when their N bodies (in the case of PFETs) connect to the highest potential possible. But since β_0 can be high, engineers sometimes activate them on purpose.

B. Substrate BJTs

Oppositely doped source–body–substrate and drain–body–substrate diffusions in wells also implement vertical and lateral *substrate BJTs*. In these cases, sources and drains supply the carriers that the substrate collects, so sources and drains are emitters, wells are bases, and the substrate is the collector. β_0 can be high because moderately doped and relatively shallow wells can establish short w_B's.

Connecting N-well bodies (in the case of PFETs) to the highest potential keeps these substrate BJTs off. Analog engineers sometimes activate them on purpose for their β_0. But this is not common practice, however, because they inject i_{SUB} into the shared substrate.

C. Substrate MOSFETs

NFETs built directly over a P substrate (like in Fig. 51) share their body with the rest of the die. So independent access to their body terminals is not possible. Engineers sometimes use three-terminal symbols (in Fig. 22) to indicate this. In the case of substrate NFETs, the P body's connection to ground or to the most negative potential is implied. *Substrate MOSFETs* incorporate the channel BJT and substrate diodes that Figs. 51 and 52 show.

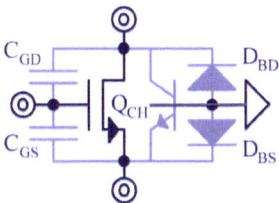

Fig. 52. Substrate NMOS with parasitic components.

D. Welled MOSFETs

The body of PFETs built in N wells over a P substrate (like in Fig. 51) is the N well. So independent access to their bodies is possible. The four-terminal symbol (in Fig. 33) is therefore more appropriate and almost always used. On occasion, engineers use three-terminal symbols to indicate a pool of welled PFETs share one well and one body connection. *Welled MOSFETs* incorporate the channel and substrate BJTs that Figs. 51 and 53 show.

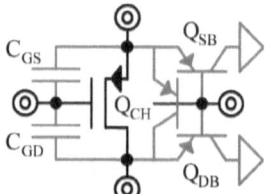

Fig. 53. Welled PMOS with parasitic components.

E. Process Variants

Although less popular, integrating N- and P-channel MOSFETs in N substrates is also possible. In these cases, PFETs sit directly over an N substrate, NFETs lie in P wells, and the N substrate connects to the most positive potential. So PFET body terminals are not available.

Independent access to the body offers a degree of design flexibility that can help optimize and improve circuit performance. Substrate MOSFETs in "vanilla" *single-well* technologies do not offer this option. *Twin-well* or *dual-well* process technologies do because they can embed NFETs and PFETs in their own independent wells. The drawback is the additional expense of more fabrication steps.

6.3. Diffused-Channel MOSFETs

Automotive, laptop, and other higher-voltage consumer products call for v_{DS} breakdown voltages that exceed the rating of 1.8–5.5-V LDD MOSFETs. *Diffused-channel MOS* (DMOS) transistors are popular in this space because they can break at 12–100 V. v_{DS} breakdown is higher because their channel–drain drift regions are longer. The drawback is the cost of the additional fabrication steps needed to build them. Still, many power supplies cannot survive without these *double-diffused* structures.

Adding an NPN P base layer to the CMOS process illustrated in Fig. 51 not only avails the vertical N^+–P base–N well BJT structure in Fig. 54 but also the *lateral DMOS* (LDMOS) under the oxide. Here, the P base is the body of the NMOS that v_G inverts. The N well extends the N^+ drain to the edge of the P-base body. So L_{CH} is the distance across the P base between the edges of the N^+ source and the N-well drain.

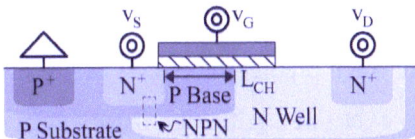

Fig. 54. Lateral diffused-channel N-channel MOSFET.

Operationally, v_G inverts the P-base region under the oxide to form an N-channel that connects the N^+ source to the N-well drain. When v_D reduces v_{GD} below v_T, the channel pinches and the region near the P base–N well junction depletes. So the saturated channel drops $v_{DS(SAT)}$'s $v_{GS} - v_T$, the non-depleted part of the N well drops an Ohmic portion of v_{DS}, and the depleted region drops the rest.

The N well is so lightly doped that it can drop 10–50 V, depending on doping concentration and t_{OX}. Unfortunately, the non-depleted portion of the N well adds on resistance to R_{CH}. So the lateral distance between the P base and N^+ drain should be as short as limitations allow.

This *drain-extended* NMOS is not the only way to realize a DMOS. Depending on the fabrication technology, *vertical DMOS* (VDMOS) transistors are also possible. Thicker oxides and V-shaped gate structures are other ways of extending the breakdown limits of the MOSFET.

6.4. Noise

Temporal noise in microelectronics refers to unintended variations in conduction. It generally falls under one of two categories: electronic or *systemic noise*. Electronic noise is innate to devices and systemic noise is the byproduct of circuit operation. In other words, electronic noise is natural and random and systemic noise is man-made and correlated.

A. Terminology

To function as prescribed, electronic systems must discern signals from noise. Signal-to-noise strength or *signal-to-noise ratio* (SNR) should be no less than five or 14 dB, and in most cases, greater than ten or 20 dB. This means that noise should be, by design, a small signal. *Small-signal models* can therefore often predict the effects of noise fairly well.

Noise manifests over time. Although hardly ever a single sinusoid, a combination of sinusoids can replicate any behavior. So decomposing noise into frequencies is convenient. *Noise spectrum* is a popular term in this respect because it refers to the frequency content of noise.

Spectral noise density n_d refers to noise power at one of these tones in W/Hz, A/√Hz, or V/√Hz. Total noise n_t is the summation of these n_d's across *operating frequencies* f_O. But since zero crossings (phase) seldom align, n_t is the statistical *root sum of squares* (RSS):

$$n_t = \sqrt{\int_{f_{LOW}}^{f_{HIGH}} n_d^2 \, df_o} \, . \tag{45}$$

Capacitances require current and time to charge and discharge. Since their voltages cannot rise or fall instantly, circuits cannot track infinitely fast signals. Capacitances also block static low-frequency components. So circuits only process signals and noise within the *frequency band* or *bandwidth* $f_{HIGH} - f_{LOW}$ or Δf_{BW} that these capacitances ultimately allow. n_t is therefore the statistical sum of n_f across this *noise bandwidth*.

B. Electronic Noise

Thermal: Temperature energizes electrons into random mobile states that cause them to collide. These collisions produce small-signal variations in current i_{nt} that intensify with temperature. Since resistivity impedes motion, the resistance R_X of the material suppresses this *thermal noise current* i_{nt}:

$$\frac{i_{nt}^2}{\int df_O} = \frac{i_{nt}^2}{\Delta f_{BW}} = \frac{4k_B T_J}{R_X}. \tag{46}$$

R_X therefore drops a *thermal noise voltage* v_{nt} that scales with T_J and R_X:

$$\frac{v_{nt}^2}{\int df_O} = \frac{(i_{nt} R_X)^2}{\Delta f_{BW}} = 4k_B T_J R_X. \tag{47}$$

These collisions are so random in nature that i_{nt} and v_{nt} decompose, like white light, into all frequencies equally. *Thermal noise* is therefore a form of *white noise*. As such, frequency strength is uniform across the spectrum and total noise is the statistical sum of strengths across Δf_{BW}.

Holes suffer from the same effect because they drift as valence electrons shift positions. The only difference is that bound valence electrons resist motion more than loosely bound electrons in the conduction band. So i_{nt} is lower and v_{nt} is higher in P material.

In MOSFETs, R_{CH} generates i_{nt} and v_{nt}. Interestingly, R_{CH} in triode inversion is equivalent to $1/g_m$ for long channels in saturated inversion:

$$R_{CH}\Big|_{v_{DS} \ll v_{GST}}^{v_{GS} > v_T} \approx \left(\frac{L}{W}\right)\left[\frac{1}{K'(v_{GS} - v_T)}\right] \approx \frac{1}{g_m}\Big|_{v_{DS} > v_{DS(SAT)}}^{v_{GS} > v_T}. \tag{48}$$

So R_{CH} is $1/g_m$ in triode and $1.5/g_m$ in saturation. Since t_{OX} is thinner with shorter L_{CH}'s, surface scattering in short-channel devices raises R_{CH} in saturation to $2/g_m$ or $3/g_m$. So R_{CH}'s $1/g_m$ to $3/g_m$ generates thermal noise.

Shot: Electrons "shoot" through gaps randomly. Diodes, BJTs, JFETs, and MOSFETs all suffer from this *shot noise* because their electrons cross the drift space that their respective depletion regions establish. The resulting *shot noise current* i_{ns} is proportional to *electronic charge* q_E and intensifies with higher conduction i_D:

$$\frac{i_{ns}^2}{\int df_O} = \frac{i_{ns}^2}{\Delta f_{BW}} = 2q_E i_D. \tag{49}$$

Shot noise is so random in nature that it spreads evenly across frequency. So like thermal noise, shot noise is also a form of white noise. i_{ns} is therefore the statistical sum of individual strengths across Δf_{BW}.

Flicker: Like the "flicker" of a flame, *flicker noise* is mostly a low-frequency phenomenon. In microelectronics, it refers to *1/f noise* because noise power falls with f_O at 20 dB per decade. It is a form of *pink noise* (from audio engineering) for this reason.

Since flicker noise n_{df} fades with f_O, white noise n_{dw} in Fig. 55 overpowers n_{df} past the *noise corner frequency* f_C where n_{df} and n_{dw} cross. f_C is an indirect measure of noise content n_d because f_C rises with n_{df}. In other words, n_{df} is more powerful when f_C is higher.

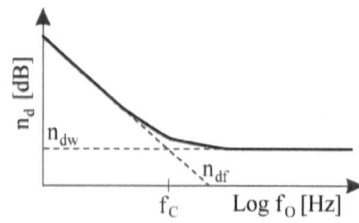

Fig. 55. Spectrum of flicker and thermal noise.

Slow carriers have more time to recombine (when crossing PN junctions and BJT bases) and fall into surface oxide traps (when crossing MOS channels) than fast carriers. Slow carriers also scatter less (along the oxide surface when crossing MOS channels) than fast carriers. So the noise that these random mechanisms produce fades with f_O.

Flicker noise in MOSFETs is usually worse because the effects of random recombination on conduction are less severe than oxide irregularities and scattering along the oxide surface. The NMOS is worse in this respect because their channel electrons are more loosely bound than the valence electrons that shift holes in the PMOS. BJTs and JFETs generate less 1/f noise because they conduct well beneath the surface of the semiconductor, where the silicon structure is less imperfect.

Surface effects in MOSFETs are more profound across shorter channels because surface defects are a larger fraction of the channel. More intense v_{GS} fields and higher conduction increases their effects. So *flicker noise current* i_{nf} climbs with lower L_{CH} and $C_{OX}"$ and higher i_D:

$$\frac{i_{nf}^2}{\int df_O} = \frac{K_F i_D}{C_{OX}" L_{CH}^2 f_O}. \tag{50}$$

Traps and their effects on scattering are so dependent on the fabrication process that engineers normally derive the *flicker noise coefficient* K_F empirically from measurements.

Side Note: *Brown noise* falls quadratically with f_O. This noise is relevant in audio engineering because low-frequency $1/f^2$ noise can overpower 1/f noise. In microelectronics, pink 1/f noise is usually more powerful, so brown $1/f^2$ noise is less discernable, and as a result, less relevant.

C. Systemic Noise

Coupling: As capacitances charge and discharge, they draw and supply *displacement currents*. When capacitances are the unintended byproducts of devices, these currents become and produce *coupled noise current* i_{nc}. i_{nc} appears everywhere because all components incorporate capacitances.

i_{nc} is systemic because the capacitances and voltages that produce i_{nc} depend entirely on the circuit. i_{nc} is therefore consistent and predictable over time. Substrate capacitances that result from junction isolation are especially problematic because they couple noise into the shared substrate. So the substrate collects, integrates, and spreads this noise to every component. Large digital blocks and switching power supplies, for example, generate *switching noise* at their *switching frequency* f_{SW} that normally appears almost everywhere in the system.

Injection: The effects of forward-biasing substrate junctions are more severe. This is because i_{SUB} can be higher and more sustained. Substrate and well resistances can therefore drop voltages that *de-bias* parasitic substrate diodes and lateral and vertical BJTs into action.

Audio amplifiers, power supplies, and power amplifiers are more prone to *injection* because they conduct lots of power at near-breakdown voltages. So impact ionization and avalanche currents are more likely to flow into and across the substrate and wells. Switched inductors in power supplies are also more likely to feed and forward-bias these junctions.

7. Summary

MOSFETs are the evolutionary offspring of JFETs. And JFETs are nothing but resistors that gate fields pinch with their depletion regions. What is perhaps most interesting is that current saturates when the channel voltage saturates. Past this $v_{DS(SAT)}$, i_D is only sensitive to v_{GS}.

MOSFETs similarly use fields to alter channel resistance. But to invert channels, v_G should overcome v_T. Their channel regions deplete below v_T and accumulate opposite charge carriers when v_G reverses.

v_G in sub-threshold reduces the barrier that carriers must overcome to diffuse, v_G in inversion pulls carriers into the channel, and v_{DS} propels all these carriers across the channel. v_D, however, opposes v_{GD}'s barrier reduction and charge formation until i_D saturates. In inversion, v_D pinches and saturates the channel voltage like JFETs. Also like JFETs, v_B is a bottom gate that can reinforce or counter the action of v_G.

Without a channel, C_{GB} incorporates all oxide capacitance to the channel region. C_{GS} and C_{GD} are low in cut off because gates overlap sources and drains across a very short L_{OL}. C_{GB}, however, loses C_{CH} to C_{GS} and C_{GD} when the channel forms. And after v_D pinches the channel, C_{GB} loses its share of C_{CH} to C_{GS} and C_{GD}, but mostly to C_{GS} because the inverted channel is a large fraction of L_{CH} that still connects to v_S.

When paralleled, these capacitances become a bi-modal varactor. Disconnecting or reversing the semiconductor type of the body eliminates the non-monotonicity that bi-modal behavior engenders. And connecting the gate and source closes a feedback loop around C_{GS} that converts the MOSFET into a diode.

Short channels are desirable in microelectronics because microchips can fit more transistors and perform more functions. Unfortunately, the depletion region that v_D induces when L_{CH} is short reaches so far into the channel that it reinforces the action of v_G. Thinning the oxide helps shunt the effect of v_D on the channel. But the stronger gate–channel and source–drain fields that result induce surface scattering, hot electron injection, oxide-surface ejections, velocity saturation, and impact ionization. Reducing v_D's doping concentration helps because it extends the short drain–channel drift region, which weakens the lateral field.

As MOSFETs invert, drift and diffusion currents compete. Luckily, switching applications switch between cut off and triode. So predicting i_D

accurately across short-lived transitions is not critical. To avoid this relatively unpredictable region of operation, analog designs often bias MOSFETs deep in sub-threshold or in strong inversion. Noting that MOSFETs effectively "invert" their small-signal gains when $v_{DS(SAT)}$ surpasses $2n_IV_t$ is helpful in this respect.

When integrated on the same substrate, substrate and welled MOSFETs incorporate substrate diodes and source–drain and substrate BJTs. Engineers zero- or reverse-bias all substrate PN junctions to de-activate unintended components and isolate designed-in devices. Diffusing a body region under the source that extends to a lightly doped N-well drain extends v_{DS}'s breakdown limit. These diffused-channel MOSFETs cost more because they require additional fabrication steps.

Thermal energy, conduction across depleted spaces, and silicon-surface imperfections produce random electronic noise in i_D. Substrate capacitances also couple and spread circuit-generated noise. The effects of near-breakdown operation are worse because impact-ionization currents de-bias substrate diodes and BJTs into action. Large circuits that switch usually suffer the most from noise coupling and injection.

www.ingramcontent.com/pod-product-compliance
Lightning Source LLC
Chambersburg PA
CBHW021901170526
45157CB00005B/1915